HAKRA

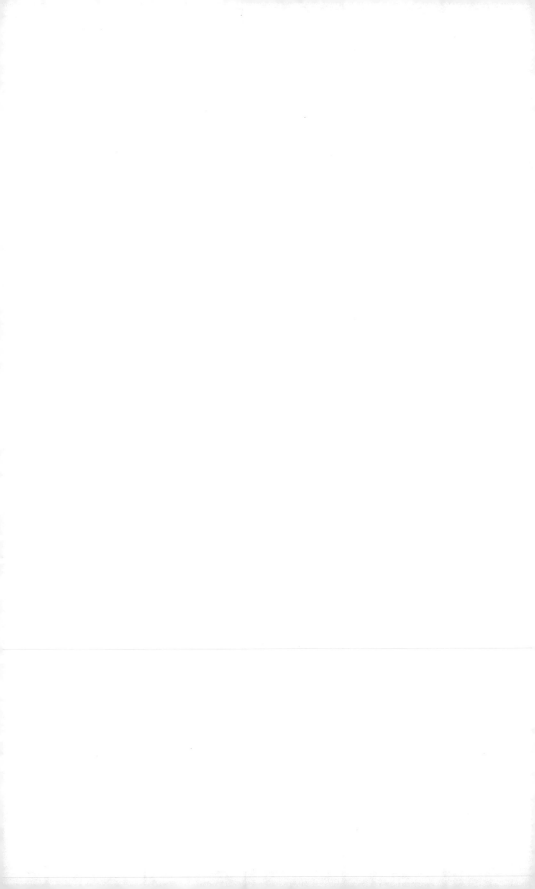

Topics in Applied Physics       Volume 74

**Springer**

*Berlin*
*Heidelberg*
*New York*
*Barcelona*
*Budapest*
*Hong Kong*
*London*
*Milan*
*Paris*
*Singapore*
*Tokyo*

# Topics in Applied Physics   Founded by Helmut K. V. Lotsch

Volume 1–48 are listed on the book inside cover

# Millimeter and Submillimeter Wave Spectroscopy of Solids

Edited by G. Grüner

With Contributions by
C. Dahl   L. Genzel   P. Goy   G. Grüner   J. P. Kotthaus
G. Kozlov   M. C. Nuss   J. Orenstein   A. Volkov

With 173 Figures

 Springer

Professor Dr. George Grüner

University of California, Los Angeles (UCLA)
Department of Physics and Astronomy
Los Angeles, CA 90024-1547
USA

ISSN 0303-4216
ISBN 3-540-62860-6  Springer-Verlag Berlin Heidelberg New York

Library of Congress Cataloging-in-Publication Data.

Millimeter and submillimeter wave spectroscopy of solids / edited by
  G. Grüner ; with contributions by K. Dahl ... [et al.].
  p.  cm. -- (Topics in applied physics ; v. 74)
Includes bibliographical references and index.
ISBN 3-540-62860-6 (alk. paper)
1. Solids--Spectra.   2. Microwave spectroscopy.   3. Submillimeter waves.
I. Grüner, George.   II. Dahl, K. (Klaus)   III. Series.
QC176.8.06.M55   1998
530.4'16--dc21                                        97-28346 CIP

© Springer-Verlag Berlin Heidelberg 1998
Printed in Germany

The use of general descriptive names, registered names, trademarks, etc. in this publication does not imply, even in the absence of a specific statement, that such names are exempt from the relevant protective laws and regulations and therefore free for general use.

Cover concept:eStudio Calamar Steinen
Cover design: *design & production* GmbH, Heidelberg
Typesetting: Protago TeX Produktion, Berlin

SPIN: 10481753        54/3140 - 5 4 3 2 1 0 - Printed on acid-free paper

# Preface

Spectroscopies that utilize the millimeter-wave spectral range of the electro-magnetic spectrum are gaining increased prominence. This is due to several factors. First, the frequency range where useful signal generation and detection becomes feasible is increasing and by now includes not only the microwave but also the millimeter- and submillimeter-wave frequencies. Second, in many fields in modern condensed-matter physics there is a clear need for the exploration of the low-energy tail of the electromagnetic response, the range where conventional optical spectroscopies become ineffective. Third, various applications also start to emerge, with devices and systems moving to higher frequencies, thus reducing the size of the hardware components.

This book discusses the various aspects of this modern and rapidly developing field. The different chapters cover the important technical developments in this area of basic and applied research. Time-domain spectroscopy is discussed along with spectroscopies performed in the frequency domain using either narrow-band, coherent or broad-band radiation sources, with Fourier transform spectroscopy as the tool of choice in the latter case. In the spectral range involving millimeter and submillimeter waves the propagation or radiation occurs either in free space or in waveguides. Techniques utilizing resonant or nonresonant measurement configurations, and also interferometric arrangements, are all standard tools used, and the advantages and disadvantages of the methods are discussed. Experiments conducted in the presence of external magnetic fields are also covered. Examples that demonstrate the utilization of the techniques include measurements on materials with known properties, but new, significant results obtained using the various techniques are also covered.

It is believed that the book will serve as an excellent source for not only readers intending to enter this developing area of science and technology, but also for those working on optical spectroscopy in other spectral ranges. Physicists in basic science and researchers and engineers will find this book equally useful and informative.

Los Angeles, January 1998                                    *George Grüner*

# Contents

# Contributors

**Dahl, Claus**
Siemens AG
Semiconductor Division
Otto-Hahn-Ring 6
D-81739 München, Germany

**Genzel, Ludwig**
Max-Plank-Institut
für Festkörperforschung
Heisenbergstrasse 1
D-70569 Stuttgart, Germany

**Goy, Philippe**
Département de Physique
de l'Ecole Normale Supérieure
24 rue Lhomond
F-75231 Paris Cedex05, France

**Grüner, George**
University of California, Los Angeles
Department of Physics and Astronomy
Los Angeles, CA 90024-1547, USA

**Kotthaus, Jörg P.**
Sektion Physik
der Ludwig-Maximilians-Universität
Geschwister-Scholl-Platz 1
D-80539 München, Germany

**Kozlov, Gennadi &**
**Volkov, Alexander**
General Physics Institute
Russian Academy of Sciences
Vavilov str. 38
117942 Moscow, Russia

**Nuss, Martin C.**
AT&T Bell Laboratories
Lucent Technologies
101 Crawfords Corner Road
Holmdel, NJ 07733-3030, USA

**Orenstein, Joseph**
Department of Physics
University of California
Berkeley, CA 94720, USA

# 1. Introduction

George Grüner
With 1 Figure

This volume summarizes the recent developments in optical spectroscopy in a well defined frequency region of the electromagnetic spectrum. The spectral range which will be discussed is somewhat loosely defined here by the frequencies where the wavelengh $\lambda$ of the radiation is one order of magnitude more or one order of magnitude less than 1 mm; the corresponding limiting frequencies are approximately 30 GHz and 3 THz. The fundamental objective (as is the case for optical methods in general) is the evaluation of the components of the complex conductivity

$$\sigma(\omega) = \sigma_1(\omega) + \sigma_2(\omega) \tag{1.1}$$

as a function of frequency. These parameters can, in turn, be compared with the calculated response, which is derived using well-known formulas and appropriate theoretical models.

The millimeter wave spectral range bridges two broad areas as far as the usually employed measurement techniques are concerned. While conventional radiofrequency methods work well up to approximately 1 GHz, above this frequency, stray capacitances and effects related to finite cable lengths, prevent the precise measurement of the components of the optical conductivity. At the upper side of the frequency spectrum discussed in this book conventional optical methods become progressively ineffective as the frequency is decreased below about 600 GHz (approximately $20\,\mathrm{cm}^{-1}$), mainly because of the decreasing source intensity, and because of problems associated with finite sample size effects. The spectral range which is the focus of this book requires a host of novel and innovative measurement techniques utilizing configurations which, in general, are not readily available on the commercial market. In the majority of cases, the development efforts which lead to such methods proceed hand in hand with advancement on the engineering side, and with applications in various areas of condensed matter physics.

The field of millimeter wave spectroscopy of solids enjoyed increased prominence in recent years due to a combination of factors. While standard optical spectroscopy played a fundamental role in conventional metals and semiconductors, in several modern areas of condensed matter physics the relevant energy scales of the various single particle and collective excitations are small. The availability of appropriate sources and detectors, together with

Topics in Applied Physics, Vol. 74
**Millimeter and Submillimeter Wave Spectroscopy of Solids** Ed.: G. Grüner
© Springer-Verlag Berlin Heidelberg 1998

the development of new measuring configurations largely contributed to the increased importance of the spectral range in modern solid state sciences.

The measurement configurations range from optical and quasi-optical arrangements, to configurations where the electromagnetic fields propagate along transmission lines, waveguides, and, occasionally, coaxial cables. The radiation sources which are used include semiconductor diodes lasers, coherent sources such as Backward Wave Oscillators and conventional broad band sources. In all cases the transmission or reflection of the electromagnetic wave is measured with the specimen inserted into the path of the electromagnetic radiation. In several cases resonant structures are also employed and these give increased sensitivity at the expense of narrow bandwidth.

Except for a few limiting cases, the components of the optical conductivity are not measured directly but have to be extracted from the measured parameters often using rather sophisticated analysis. The problem at hand is the following: At low frequencies where the wavelength of the radiation and the skin depth are significantly larger than the dimensions of the specimen, the electromagnetic field is uniform throughout the sample, which then can be represented as a combination of simple resistance $R$ and capacitance $C$. The components of the conductivity are then simply related to the frequency dependent response, as given by the $RC$ circuit. At the other end of the frequency range of interest here, at high frequencies where the wavelengh of the radiation is significantly smaller that the dimensions of the specimen, the latter can be regarded as a infinite medium – at least in two directions, usually perpendicular to the impeding electromagnetic field. This allows a straightforward analysis of the (conventional) optical constants such as the reflectivity $R(\omega)$ and transmission coefficient $T(\omega)$. In the frequency range discussed in this book these simplifying limits do not apply for several reasons. First, the sample dimensions are often between $1\,\mathrm{mm}$ and $1\,\mathrm{cm}$, comparable to the wavelength. Second, the skin depth

$$\delta = \left(\frac{\mu_0 \omega}{\sigma}\right)^{\frac{1}{2}}, \tag{1.2}$$

where $\mu_0$ is the permeability of free space and $\omega = 2\pi f$ is the measurement frequency for a moderately conducting solid is comparable to the typical sample dimension. All this requires special care when the measured parameters (such as attenuation constants, phase shifts, quality factors, line impedances, etc.) are related to the components of the optical conductivity. Broadly speaking, two limits are of importance; the so-called surface impedance limit and the so-called depolarization limit. In the former, the propagation of the elecromagnetic radiation is limited to a surface layer, given by $\delta$; in the latter, the electromagnetic field, while modified by depolarization factors, fully penetrates the specimen. The relation between the conductivity and measured parameters is, as expected, fundamentally different in the two cases. The analysis of the data and the evaluation of the components of the complex conductivity often involves considerations on finite size effects and relies upon the solution of complicated boundary problems. This important aspect of the field is discussed at length in the various chapters.

In spite of these complications, when properly applied, these techniques lead to unique information, not available by other techniques, on the physics at hand in several modern fields of condensed matter physics. The objective is achieved by employing the various measurement techniques – developed by the authors of this book, who played an important role in advancing the field, and most often made pathfinding contributions in technique development, analysis, and applications. Subsequently, these techniques were used by the authors and also by other groups in various areas of solid state physics and materials science.

The chapter by *M. Nuss* and *J. Orenstein* discusses the technique commonly known as Time Domain Spectroscopy (TDS). The method is less than a decade old, and is based on the availability of highly coherent radiation in pulse form of extremely short duration. Imaginative optoelectronic configurations allow the generation and detection of pulse waveforms propagated in the medium which includes the specimen to be measured. The frequency dependent dielectric properties of the sample can be extracted from the waveform by appropriate conversion from time-dependent to frequency-dependent response functions. The method has been used to study the properties of a wide range of solids, from semiconductors to superconductors.

The chapter by *G. Grüner* summarizes the various measurement methods which utilize waveguide configurations, the method can be called Waveguide Configuration Spectroscopy (WCS). Both resonant and nonresonant techniques are employed, and both have distinct advantages and disadvantages. Resonant techniques, where the specimen forms part of the resonant configuration, offer extreme sensitivity by enhancing the interaction between the specimen and the electromagnetic radiation. The quality factor $Q$ and the resonant frequency $f_0$ are measured and these two parameters allow for the simultaneous evaluation of both components of the conductivity. The obvious disadvantage is the narrow bandwidth as such methods operate at a single frequency, with different structures required for each frequency desired to overcome this problem tunable cavities can be employed, or one can use different modes so that different structures are not required for measurements conducted at different frequencies. Nonresonant (usually bridge) configurations can be employed for moderately conducting solids; here the attenuation $A$ and phase shift $\phi$ are measured with the sample placed into the waveguide, again using these two parameters as the input allows $\sigma_1(\omega)$ an $\sigma_2(\omega)$ to be evaluated. The extreme sensitivity of the methods has been demonstrated by experiments on materials with known electrodynamical response, and the technique has been applied in various areas of correlated solids.

The chapter by *G. Kozlov* and *A. Volkov* summarizes the technique which utilizes powerful coherent radiation sources called Backward Wave Oscillators which cover the entire millimeter wave spectrum, and the name Coherent Source Spectroscopy (CSS) is appropriate. Quasi-optical configurations are employed here, and the method has been applied both in transmission and reflection configurations. The components of the conductivity are extracted by an (often elaborate) analysis. The method has been used succesfully to study the spectral features of unconventional insulators, dielectrics, and, recently, to explore the deviations from simple metallic behavior in various conducting materials.

Fourier transform spectroscopy (FTS) played an especially important role in the optical spectroscopy of solids. The chapter by *L. Genzel* summarizes the latest developments, which allow experiments to be extended in the millimeter wave range. The method has been used, among other topics, to explore the unusual properties of the recently discovered oxide superconductors, a field where extremely sensitive measurements are required in order to elucidate the electrodynamics and the nature of the ground state.

The measurement of the dynamical conductivity in the presence of an external magnetic field is addressed by the contribution of *C. Dahl, P. Goy,* and *J.P. Kotthaus.* Special emphasis is put on the broad band magneto-transmission spectroscopy of low dimensional electron systems in semiconductors. Such studies are made possible with unprecedented ease utilizing an innovative vector analyzer operating at frequencies between 8 and 800 GHz. The main focus of this chapter lies on the millimeter wave spectroscopy of magnetoplasma resonances in laterally confined two-dimensional electron systems of different geometries.

In Fig. 1.1 the various experimental techniques which are discussed in this book are summarized together with the frequency regions covered by these techniques.

The various methods played an important, and often dominant role in exploring the various excitations of a broad range of solids, and below I list just a few examples.

While in simple metals, such as copper, the single particle energies, the plasma frequency, the single particle bandwidth and relaxation rate are of the order from 1 to 10 eV (corresponding to the UV spectral range), in several correlated metals these energies are significantly reduced. The response is given by a renormalized plasma frequency, reflecting a renormalized effective mass, and by a renormalized relaxation time which, in general, falls in the millimeter wave spectral range. Such renormalization effects have been explored in various so-called "heavy fermion" materials, and by now the relation between thermodynamics and electrodynamics is well established in these novel metallic systems. The low energy behavior of various interacting and low-dimensional solids have also attracted recent interest, mainly because of questions related to the applicability of Fermi liquid theory in these systems. Millimeter wave spectroscopy is expected to play an important role in answering these yet unresolved questions.

The single particle gaps $\Delta$ of various old superconductors such as Nb, and recently discovered materials with the transition temperature of the order of 10 K also fall in the millimeter wave range, and aside from establishing the

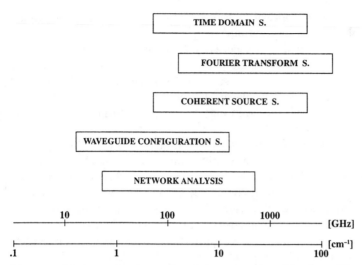

**Fig. 1.1.** Various experimental techniques employed in the millimeter-wave spectral range and the frequency range in which they are commonly employed. Each technique is discussed in separate chapters of the book

superconducting gaps and unusual phenomena known as superconducting coherence factors, the methods have been succesfully employed to measure other parameters of the superconducting state such as the penetration depth and conductivity. These experiments led to crucial new information of the superconducting state in many materials, including questions concerning the symmetry of the superconducting ground state.

Other broken symmetry ground states of metals, such as charge and spin density wave states also display a variety of charge excitations (both single particle and collective). Many of these excitations, such as the oscillations of the condensate which is pinned to the underlying lattice, (occuring at the so-called pinning frequency $\omega_0$), bound state resonances together with the single particle gaps $\Delta$ occur in the millimeter wave spectral range. Due to careful experiments on a range of model compounds, our understanding of the excitations of these novel states of matter is nearly complete. In certain superconductors experiments well below the gap frequeny lead to important information on the dynamics of the vortex state.

Metal-insulator transitions, brought about by localization effects as one parameter such as the composition is varied (the well-known Anderson localization) are characterized by a divergent dielectric constant and collapse of the excitation spectrum at the transition. The progressive development of localized states can therefore be followed by extending the frequency range of the experiments from optical to millimeter wave frequencies and below.

Similar frequency collapse occurs in the case of various phase transitions where the so-called Goldstone mode (with frequency going to zero at the transition temperature) is associated with charge degrees of freedom, and therefore is accessible by optical methods. Certain structural phase transitions are typical examples of such phenomena.

The various energy scales, associated with mesoscopic systems are significantly smaller than typical single particle energies of solids. In laterally confined two-dimensional electron systems with a typical lateral size of 10 micrometers, the charactersistic plasma frequency is of a few 100 GHz and falls in the millimeter wave spectral range, making the methods discussed in this book particularly useful.

The chapters in this book focus only on certain aspects of millimeter wave spectroscopy with applications in a limited number of areas. The field which enjoyed enormous growth in the last decade, is certainly in its infancy. With new active and passive components being developed, with novel measurement techniques appearing, and with increrasing emphasis on low energy phenomena in many modern areas of condensed matter science, the spectral range will undoubtedly play an increasingly important role in condensed matter physics, and, eventually also in materials characterization and testing.

# 2. Terahertz Time-Domain Spectroscopy

Martin C. Nuss and Joseph Orenstein
With 25 Figures

The terahertz, or far-infrared, region of the electromagnetic spectrum is of critical importance in the spectroscopy of condensed matter systems. The electronic properties of semiconductors and metals are greatly influenced by bound states (e.g., excitons and Cooper pairs) whose energies are resonant with terahertz photons. The terahertz regime also coincides with the rates of inelastic processes in solids, such as tunneling and quasiparticle scattering. As a final example, confinement energies in artificially synthesized nanostructures, like quantum wells, lie in the terahertz regime.

In spite of its importance, terahertz spectroscopy has been hindered by the lack of suitable tools. As pointed out in the introduction to this book, swept-frequency synthesizers for millimeter- and submillimeter-waves are limited to below roughly 100 GHz, with higher frequencies only available using discrete frequency sources. Fourier Transform InfraRed (FTIR) spectroscopy, on the other hand, is hampered by the lack of brightness of incoherent sources. In addition, FTIR methods are not useful if the real and imaginary part of response functions must be measured at each frequency.

Terahertz Time-Domain Spectroscopy (THz-TDS) is a new spectroscopic technique that overcomes these difficulties in a radical way. Its advantages have resulted in rapid proliferation within the last few years from a handful of ultrafast laser experts to researchers in a wide range of disciplines. THz-TDS is based on electromagnetic transients generated opto-electronically with the help of femtosecond (1fs = $10^{-15}$ s) duration laser pulses. These THz transients are single-cycle bursts of electromagnetic radiation of typically less than 1 ps duration. Their spectral density spans the range from below 100 GHz to more than 5 THz. Optically-gated detection allows direct measurement of the terahertz electric field with a time resolution of a fraction of a picosecond. From this measurement both the real and imaginary part of the dielectric function of a medium may be extracted, without having to resort to the Kramers-Kronig relations. Furthermore, the brightness of the THz transients exceeds that of conventional thermal sources, and the gated detection is orders of magnitude more sensitive than bolometric detection.

Recent developments have shown that THz-TDS has capabilities that go far beyond linear far-infrared spectroscopy. Because the THz transients are perfectly time-synchronized with the optical pulses that generate them, THz-TDS is ideally suited for "visible-pump, THz-probe" experiments. In

Topics in Applied Physics, Vol. 74
**Millimeter and Submillimeter Wave Spectroscopy of Solids** Ed.: G. Grüner
© Springer-Verlag Berlin Heidelberg 1998

these measurements, the optical pulse is used to optically excite a sample and the THz pulse probes the change in dielectric function (or conductivity) as a function of time after optical excitation. Another powerful nonlinear technique is the detection of THz emission following pulsed laser excitation. In many cases, these nonlinear techniques are proving even more powerful than linear THz-TDS spectroscopy and yield complementary information that may otherwise be difficult to obtain.

Beyond the characterization of new materials and the study of basic physical phenomena, the impact of THz spectroscopy in the commercial world is growing. Promising applications include industrial process control, contamination measurement, chemical analysis, wafer characterization, remote sensing, and environmental sensing. Although most of these issues are outside the solid-state focus of this book, they are instrumental for the development of the THz-TDS technology.

In light of the rapid expansion and development of THz-TDS and ultrafast lasers, there is a need for an updated review of the techniques involved. This chapter is intended as a "How-To" for researchers who wish to venture in this field as well as for those who dare to invest in commercializing this new and promising technology. We focus mainly on experimental methods: generation and detection of THz transients, different experimental methods, and the analysis of time-domain data to yield frequency-dependent dielectric functions. Rather than discussing the vast number of experimental results using THz-TDS in detail, we supply references to allow readers to follow up on their own interests. Although we have tried to be complete in referencing different experiments, there are bound to be inadvertent omissions, for which we apologize in advance.

## 2.1 Historical Development

THz-TDS is barely 15 years old. Unlike other spectroscopic techniques that have evolved more incrementally, we can clearly identify the technological breakthroughs that have enabled the technique to flourish. Femtosecond laser sources have become widely available commercially within the last few years and can now be operated by a relative novice in laser technology. Developments in materials, such as low temperature GaAs, electro-optic crystals, poled-polymers doped with chromophores, etc., have helped to unify optics and electronics. Advances in micro- and nanofabrication allow the propagation of optoelectronic signals without loss of bandwidth.

The union of these enabling optical and electronic technologies is commonly referred to as "ultrafast optoelectronics" [2.1]. The historical development of THz-TDS is intimately tied to that of ultrafast optoelectronics. Initially, electrical pulses were generated and detected on transmission lines using photoconductors excited by laser pulses [2.2]. Picosecond microwave pulses in free space were first generated by coupling these electrical pulses to a microwave antenna [2.3]. Simultaneously with the development of shorter and shorter laser pulses, advances in VLSI lithographic techniques allowed fabrication of smaller radiating structures and, consequently, higher frequency electromagnetic radiation [2.4–6]. This development culminated in the generation of terahertz bandwidth single-cycle pulses by photoconducting dipole antennas [2.7, 8]. Although these devices are broadband, they are marginally useful because they radiate isotropically. A breakthrough came with the addition of substrate lenses to couple the radiation in and out of the photoconducting dipole chip more directionally and efficiently [2.9]. This opened the way to "Terahertz Beams" [2.10], which could be collimated and focused as easily as light beams in a spectrometer.

It is interesting to note that prior to the development of THz bandwidth photoconductive dipole antennas, THz electromagnetic transients were generated by optical rectification in electro-optic crystals like $LiTaO_3$, which also allowed for simultaneous electro-optic detection of the transients in the same crystal [2.11–13]. Because of the difficulty in coupling the THz radiation in and out of the electro-optic crystals, this technique has now been displaced by the photoconductive antenna approach.

## 2.2 THz Time-Domain Spectrometers

We turn now to a description of THz-TDS methods applied to linear spectroscopy. Figure 2.1 is a schematic diagram of a THz-TDS spectrometer. It consists of a femtosecond laser source (1), a beam splitter that divides the laser beam into two, an optically-gated THz transmitter (2), focusing and

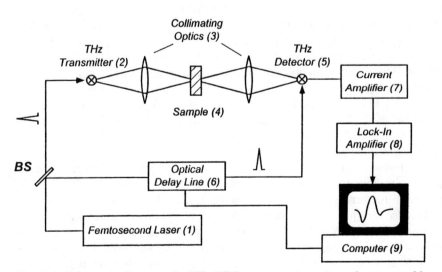

**Fig. 2.1.** Schematic diagram of a THz-TDS spectrometer using a femtosecond laser source and photoconductive THz transmitters and receivers

collimating optics (3), the sample (4), an optically-gated THz detector (5), a variable delay line (6) that varies the optical delay between the pulses gating the THz transmitter and detector, a current amplifier (7), and a lock-in amplifier (8). A computer (9) controls the variable delay line and displays the detector photocurrent versus path length. The function of each of these components is discussed in the following paragraphs.

### 2.2.1 Lasers

In the past, only dye lasers were capable of generating optical pulses of roughly 100 fs duration [2.14]. Today, solid-state lasers like the Ti:Sapphire laser [2.15, 16], delivering pulses with a wavelength near 800 nm, have largely displaced dye lasers and are commercially available. The typical repetition rate of these lasers is about 100 MHz. Not more than 10–50 mW of laser power is required for either THz transmitter or detector. In the foreseeable future, progress in diode-pumped solid-state laser sources [2.17–19] will make femtosecond lasers more practical, compact, and cost-effective. Recent developments promise that a complete THz time-domain spectrometer will soon fit into a smaller box than a commercial FTIR spectrometer.

### 2.2.2 THz Transmitters and Detectors

The truly elegant aspect of the THz time-domain linear spectrometer is that the sources and detectors consist of the same building block. Both are based on the photoconductive (Auston) switch [2.20, 21], which consists of a semiconductor bridging the gap in a transmission line structure deposited on the semiconductor substrate (Fig. 2.2a). The response of the voltage-biased photoconductive switch to a short optical pulse focused onto the gap between the two contacts is illustrated in Fig. 2.2b. The current through the switch rises very rapidly after injection of photocarriers by the optical pulse, and then decays with a time constant given by the carrier lifetime of the semiconductor. The transient photocurrent $J(t)$ radiates into free space according to Maxwell's equations, $E(t) \propto \partial J(t)/\partial t$. Because of the time derivative, the radiated field is dominated by the rising edge of the photocurrent transient, which is invariably much faster than the decay. Long tails of the photocurrent decay, which occur in most semiconductors without high defect density, are largely irrelevant to the radiated field.

To convert the Auston switch for use as a detector of short electrical pulses, an ammeter (or current-to-voltage amplifier) is connected across the photoconductor, replacing the voltage bias. The electric field of an incident THz pulse now provides the driving field for the photocarriers. Current flows through the switch only when both the THz field and photocarriers are present. Since electronics are not fast enough to measure the THz transients directly, repetitive photoconductive sampling is used (Fig. 2.3). If the photocarrier lifetime, $\tau$, is much shorter than the THz pulse, the photoconductive

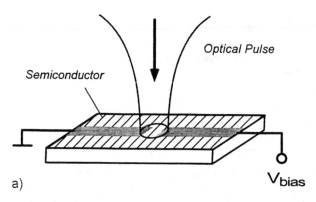

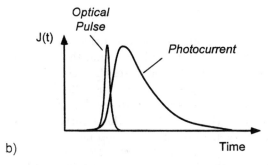

**Fig. 2.2.** (a) A photoconductive "Auston" switch integrated in a transmission line. Optical pulses are focused onto the gap in the transmission line and inject carriers into the semiconductor, leading to a current transient $J(t)$ flowing across the gap in the line. (b) Typical current response $J(t)$ of a photoconductive switch to a short optical excitation pulse

switch acts as a sampling gate which samples the THz field within a time $\tau$. Because the laser pulses which trigger the transmitter and gate the detector originate from the same source, the photoconductive gate can be moved across the THz waveform with an optical delay line (Fig. 2.1). Using this technique, the entire THz transient is mapped without the need for fast electronics.

**a) Transmitter Structures.** While receiver and detector structures were initially identical [2.8, 22], more efficient transmitter structures have since been devised. Figure 2.4 shows the design of a high-efficiency transmitter structure [2.23]. The structure consists of two $\approx 10\,\mu$m wide metal lines deposited on a semi-insulating GaAs wafer, separated by $\approx 50$–$100\,\mu$m. These two lines form Schottky diodes on top of the GaAs substrate. A voltage of about $100\,$V is applied across the two lines, creating a very strong depletion field near the positive (anode) contact, which becomes reverse biased. This strong field is enhanced by the local field from deep traps present in semi-insulating wafers [2.24]. Femtosecond laser pulses are focused to a $\approx 10\,\mu$m

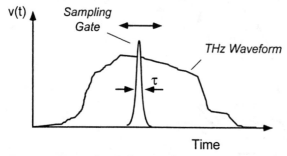

**Fig. 2.3.** Principle of photoconductive sampling. The photoconductive switch acts as a sampling gate that measures the waveform voltage $v(t)$ within the sampling time $\tau$. By changing the delay between the optical pulse triggering the sampling gate and the waveform, the entire waveform can be mapped out sequentially in time

diameter spot close to the anode, generating photocarriers that are accelerated at an enormous rate in the large field. It has been found that focusing the laser beam close to electrode corners further increases the THz radiation because of the well-known field enhancements that occur there [2.25, 26].

The availability of amplified optical pulses has led to the development of a very different transmitter structure (Fig. 2.5) [2.27]. This structure differs from the Auston-switch based THz generator in two important ways. First, the illuminated area is quite large, typically in the range from $mm^2$ up to $1\,cm^2$ [2.28, 29]. This is in contrast with generation using unamplified pulses, where the laser beam is focused to a spot of order $10\,\mu m$. The THz generation remains efficient because the optical intensity is high enough to reduce the photoconductor's impedance to approximately the free-space impedance over the entire illuminated area. Second, the antenna structure is no longer required. Because the illuminated area is large compared with the THz wavelength the THz radiation emerges colinearly with the reflected laser beam. This can be regarded as the phased array-like action of each illuminated element of the semiconductor surface [2.27].

Because of the built-in surface field, THz emission can be observed from the bare semiconductor surface without applying an external field [2.30]. An

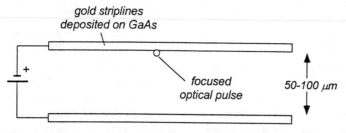

**Fig. 2.4.** Structure of a high-efficiency GaAs THz transmitter. The optical pulses are focused near the positive electrode on the GaAs wafer

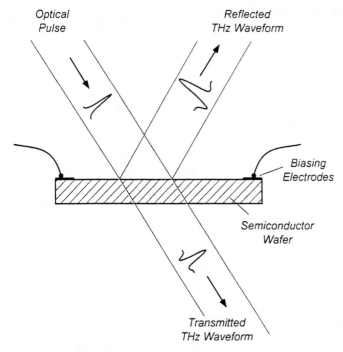

**Fig. 2.5.** Large-aperture geometry for generating THz transients. THz pulses emerge from the semiconductor wafer both colinearly with the reflected optical pulse, as well as from the back of the wafer in the same direction as the incident optical pulse

applied electric field greatly enhances the THz emission, however. Terahertz electric fields as high as $10^5$ V/cm at the beam waist have been reported [2.29]. The static fields may be applied parallel to the surface using electrodes [2.27, 31], or perpendicular to the surface using p-i-n diode structures [2.32] or a Schottky metal contact and a conducting substrate [2.33]. In general, there is also a THz beam emitted into the substrate and emerging from the backside of the semiconductor wafer (Fig. 2.5) [2.30].

The availability of high-intensity THz radiation from large-area transmitters is opening new fields of research. "Half-cycle" terahertz radiation has been used to prepare Rydberg atoms in unusual excited states where, for example, only one side of the atom is ionized. These exotic states are possible because the terahertz pulse creates a large distortion of the atomic potential on a time scale which is short compared to traversal time of a Rydberg wavepacket. [2.34] In addition to atomic physics, we expect that high-intensity terahertz radiation will generate new nonlinear effects in condensed matter systems, as well.

**b) Speed of THz Transmitter.** As carriers are created by absorption of photons nearly instantaneously on the time scale of interest, the speed of the radiated electromagnetic transient is limited by: (i) how fast the carriers can be accelerated to create a current transient, and (ii), the impulse response of the radiating structure. The dynamics of photoexcited electron-hole pairs in a large field has been discussed theoretically and experimentally by many researchers. Their motion is determined by the band structure of the semiconductor, the momentum scattering rate, and the size of the electric field. The band structure is particularly important in GaAs, where velocity saturation can occur on a subpicosecond timescale due to intervalley transfer of the electrons into low-mobility satellite valleys. If carriers are injected close to the satellite valleys, this can slow the photocurrent risetime to about 2 ps [2.12], compared to 200 fs measured for carriers injected close to the band edge [2.35]. The latter generates electromagnetic transients with bandwidth in excess of 4 THz, in the absence of band-narrowing from the radiating structure [2.36].

It is important to note that free carrier acceleration is not a fundamental limit to the bandwidth of THz transients, as the above arguments only apply to the transport current, not the displacement current. It has been shown that the THz radiation from the displacement current can be comparable to that from the transport current [2.37]. A physical picture for the displacement current is that in the presence of a field, electrons and holes are excited into states that are spatially displaced from each other. This results in an "instantaneous" polarization of the e-h pairs upon creation [2.38], whose time derivative corresponds to a current that is as fast as the optical pulse that generates the carriers. In quantum wells at low temperature, this is the only contribution to the current when the applied field is perpendicular to the wells [2.37]. In bulk semiconductors, the displacement component dominates for photoexcitation below the bandgap [2.39, 40]. The mechanism may also be regarded as optical rectification through the nonlinear susceptibility $\chi^{(2)}(0; \omega, -\omega)$. The second order susceptibility is allowed by either the symmetry breaking effect of the applied electric field or the semiconductor surface itself. Recently, it has been shown that this instantaneous polarization can indeed emit THz radiation that extends out to the mid-IR frequency range [2.41].

Bandwidth narrowing due to the radiating structure can be largely eliminated by scaling the structures to sizes much smaller than the wavelength [2.7, 8]. It is also possible to eliminate the dipole structure altogether [2.7, 23, 27], in which case the purpose of electrical contacts is only to apply the bias field. Then, the transmitter resembles a point source (Hertzian) dipole transmitter, with the frequency response only determined by the first time derivative of the current. In some situations the built-in surface field may be used as a bias field, resulting in THz radiation from bare semiconductor wafers [2.30].

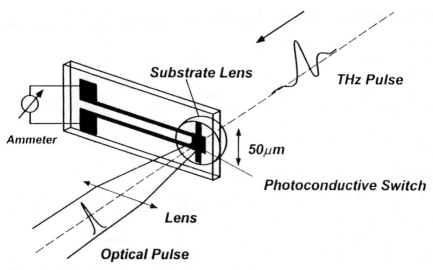

**Fig. 2.6.** Photoconducting dipole antenna [2.8]

**c) Detector Structure.** Figure 2.6 shows a photoconducting dipole antenna fabricated on radiation-damaged silicon-on sapphire (RD-SOS) [2.8]. The total dimension of the dipole arm is 50 μm, with a gap between the two dipole arms of 5 μm. After evaporating the antenna, the typically 0.6 μm thick silicon layer is etched away except for a 100 × 100 μm square directly underneath the dipole. This step increases the dark resistance and reduces the dark current of the antenna, which can be a source of noise for the detector. Structures with feed lines on the outside rather than at the center of the dipole have also been used successfully [2.22]. For dipole lengths larger than 100 μm, these antennas behave like resonant dipoles. For smaller sizes, the structure behaves like an elementary Hertzian dipole, and the bandwidth of the current response rather than the dipole, determines the frequency response [2.8]. Figure 2.6 also shows a substrate lens that is commonly attached to the substrate to improve the coupling of the THz pulses in and out of the substrate.

**d) Gated Detection.** In the dark, the photoconductive gap of the detector antenna is highly resistive ($\approx 20\,\mathrm{M\Omega}$). Injection of carriers by the laser pulse causes the resistance to drop below 500 Ω. During the photocarrier lifetime a current flows which is proportional to the amplitude of the received THz field. The current is converted to a voltage by a current amplifier connected across the feed lines of the antenna.

Varying the time-delay between the THz transient and the optical gating pulse maps out the entire THz waveform (Fig. 2.3). There are a number of ways in which this measurement can be performed. In the most common, the optical beam exciting the transmitter is mechanically chopped and the voltage from the current amplifier is synchronously detected using a lock-in amplifier. The optical delay is slowly scanned and the photocurrent acquired into a computer. Another technique is "rapid-scan", in which the time-delay is scanned at a rate of tens to hundreds of Hz using a shaker with an optical retro-reflector. To enhance the signal-to-noise ratio, each scan is co-added using an averaging digital oscilloscope. Rapid-scan can significantly reduce the noise due to 1/f laser power fluctuations. In many applications, the photocurrent signal is so large (nA level), that the output from the current amplifier can be directly digitized for further processing without using a lock-in amplifier [2.35].

**e) Frequency Limit of THz Detectors.** The bandwidth of the detection process is determined by two factors, the photocurrent response and the frequency-dependence of the antenna structure. In general, the low-frequency cutoff of the detectors results from the collection efficiency of the dipole, while the upper frequency limit is determined by the photocarrier response. We focus first on the photocurrent response, which is the convolution of the transient photoconductivity $\sigma(t)$ and the electric field $E(t)$ across the photoconductor:

$$J(t) = \int \sigma(t - t') \cdot E(t')dt' \tag{2.1}$$

where $J(t)$ is the photocurrent transient. $E(t)$ is faithfully reproduced by $J(t)$ when the photocurrent transient becomes much shorter than the THz waveform.

The photocurrent decay time in the Auston switch must be less than roughly 0.5 ps in order to resolve transients in the THz regime. Recombination in semiconductors with low defect density tends to be far slower, therefore the carrier lifetime has to be reduced below its intrinsic value. This is commonly accomplished by introducing defect states that have a fast carrier capture rate, either during crystal growth or afterwards by ion implantation. An example of the first case is low temperature grown GaAs (LT-GaAs), which has been shown to have carrier lifetimes as short as 280 fs when properly annealed [2.42]. An example of the latter case is radiation-damaged silicon-on-sapphire (RD-SOS), in which dislocations are formed by implanting argon, silicon, or oxygen ions [2.8, 43].

The electric field across the photoconductor can differ from the THz pulse in free space due to the frequency-response of the antenna structure. Using the reciprocity principle, the collection efficiency of the detector is identical to the radiation efficiency of the transmitter. For a Hertzian dipole, where the antenna dimension is much less that the wavelength, the radiation efficiency (and thus the collection efficiency) is proportional to $\omega$ (corresponding to the first derivative of the current). For "real" dipoles, the frequency response will be more complicated.

It should be noted that there are applications in which electromagnetic emission and detection at frequencies below the THz range is desired. In this case, there are a number of antenna designs like the Vivaldi [2.5, 6], the bow-tie, and the log-spiral antenna [2.44], that deliver better transmission and receiver efficiency than a Hertzian dipole over a rather wide frequency range, but are typically best for frequencies below 1 THz.

**f) Radiation Pattern and Substrate Lenses.** The angular dependence of the radiation (and by reciprocity also the reception) differs from a dipole in free space because of the presence of the substrate on one side of the antenna structure. Figure 2.7 shows the calculated radiation pattern of a dipole antenna on a substrate of dielectric constant $\varepsilon = 12$ [2.45–47]. The pattern is more complicated than the $\sin^2 \theta$ dependence expected for a simple dipole in a vacuum due to the presence of the substrate dielectric. The radiation is concentrated in the dielectric and is collimated more strongly. This effect has been verified experimentally for THz dipole antennas [2.8].

An important element of both transmitter and receiver dipole antennas is the substrate lens. Without this lens the coupling into free-space is limited by the excitation of slab modes between the substrate surfaces [2.8]. Also, the substrate lens increases the collimation of the emitted electromagnetic radiation. Finally, the lens serves to magnify the dipole antenna and thus increases its efficiency [2.45]. The dielectric constant of the substrate lens should match that of the substrate in order to minimize reflections at the substrate/lens interface. In the case of GaAs, sapphire, and silicon substrates, high resistivity silicon lenses are best suited because of their low THz absorption, frequency-independent refractive index, cubic crystal structure, and ease of cutting and polishing.

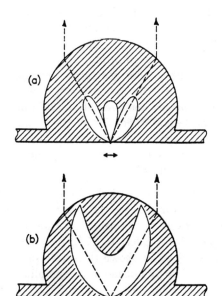

**Fig. 2.7 a,b.** Calculated power radiation pattern of a dipole located at the interface of a substrate with a dielectric constant $\varepsilon = 10$ and air ($\varepsilon = 1$) [2.46]. The dipole is oriented parallel to the interface, and (a) and (b) are for the dipole oscillating in and out of the plane of the paper, respectively

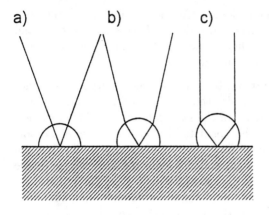

**Fig. 2.8 a–c.** Three different substrate lens designs. (**a**) the non-focusing hemispherical design, (**b**) the hyper-hemispherical lens, and (**c**), the collimating substrate lens

Three possible substrate lens designs are depicted in Fig. 2.8. The size of the lenses is usually between 2 and 10 mm in diameter. In the hemispherical design (Fig. 2.8a), the dipole source is located at the center of the lens and all rays exit the lens/air interface at normal incidence. The radiation pattern inside the substrate is preserved in the transition to free-space because no refraction occurs. The design in Fig. 2.8b is the aplanatic hyper-hemispherical substrate lens [2.45, 48]. Like the hemispherical design, it has no spherical aberration or coma, and when using silicon as a lens material, no chromatic dispersion. The design specification for the aplanatic design is that:

$$\rho = \frac{r}{n},\tag{2.2}$$

where $\rho$ is the distance from the center of the lens to the focal point, $r$ the radius of the lens, and $n$ its refractive index [2.48]. This design provides a slight collimation of the beam, which allows the remaining optical system to be designed with higher $f$-number optics. The design in Fig. 2.8c locates the dipole source at the focal point of the lens so that geometrical optics predicts a fully collimated output beam. However, diffraction is important because of the long free-space wavelength of the terahertz radiation (300 μm in air for 1 THz radiation) compared to the lateral dimension of the beam exiting the lens, and, unless substrate lenses with a diameter much larger than the wavelength are used, the beam propagates with a larger (frequency-dependent) divergence compared to the other designs.

**g) Other THz Detectors.** Besides photoconductive switches, other processes may be used to detect THz transients. Examples are the electro-optic effect in crystals like lithium niobate [2.13], excitonic electro-absorption in GaAs [2.49, 50], and the Quantum-Confined Stark Effect (QCSE) in quantum wells. However, none of these processes has so far shown the ease-of-use of the photoconductive technique. Moreover, while photoconductive sampling is background-free, these methods usually sense a small change superimposed on a large signal. In spite of the difficulties, these processes may prove useful in applications where the ultimate bandwidth is required.

### 2.2.3 Collimating and Focusing Optics

To perform spectroscopic measurements, the THz transmitter and receiver are incorporated into an optical system. The optical system must guide the radiation from source to detector, and focus the radiation to a diffraction-limited spot on a potentially small sample.

Such an optical system is shown in Fig. 2.9. Off-axis paraboloid mirrors are used to collimate and focus the THz radiation. In the system shown in Fig. 2.9, the full emission angle of the terahertz radiation emerging from the hyperhemispherical substrate lenses is about 30 degrees. Using off-axis paraboloids with 6.6 cm focal length, the far-infrared radiation is collimated to a parallel, diffraction-limited beam of roughly 25 mm diameter. This beam can be focused to a spot size of 0.5 mm at a peak frequency of 1 THz by inserting a pair of focusing lenses or off-axis paraboloids into the system (inset in Fig. 2.9).

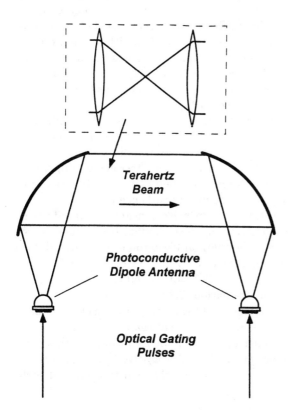

**Terahertz Beam**

**Photoconductive Dipole Antenna**

**Optical Gating Pulses**

**Fig. 2.9.** Optical setup of a THz-TDS spectrometer. The THz radiation is coupled in and out of the photoconducting dipole antenna using substrate lenses, and collimated to a parallel and diffraction-limited beam with off-axis paraboloid mirrors. Optionally, a pair of lenses or paraboloids may be used to focus the THz beam to a diffraction-limited spot at the sample (*dashed box*)

Although the paraboloids are somewhat difficult to align, they offer high reflectivity and achromatic operation over the entire THz range. As an alternative, fused quartz lenses may be used at frequencies below 1 THz, and silicon lenses up to 10 THz. Unfortunately, optical alignment with visible laser beams is impossible with fused quartz lenses because the refractive index is very different at visible and THz frequencies. Another useful lens material is TPX (poly-4-methyl-pentene-1), a polymer which has low absorption and dispersion throughout the THz range [2.51], but is difficult to polish because of its softness. Some groups have dispensed with optics beyond the substrate lens and use large diameter (25 mm) substrate lenses of the design shown in Fig. 2.8c to form a collimated beam [2.52].

## 2.3 THz Time-Domain Linear Spectroscopy: Characteristics and Advantages

### 2.3.1 Signal to Noise Ratio and Dynamic Range

A THz waveform measured with the spectrometer shown above, in absence of a sample, is shown in Fig. 2.10. The source and detector are a GaAs transmitter chip biased at 100 V and an LT-GaAs receiver chip with a 50 µm dipole, respectively [2.25]. Silicon hyperhemispherical microlenses are attached to both transmitter and receiver. The measured peak photocurrent is 270 nA, and the signal-to-noise ratio is nearly $10^6$ when using a 300 ms time constant on the lock-in amplifier (see inset in Fig. 2.10).

*Van Exter* and *Grischkowsky* [2.22] has estimated the average power of the THz beam at $\approx 10$ nW, in a very similar spectrometer. The peak power is much higher, by a factor $10^4$, because the energy appears in 1 ps bursts every 10 ns. The energy per burst is $\approx 0.1$ fJ, corresponding to roughly 50 000 THz photons. The reason for the large S/N ratios is the use of gated detection, the detector is off for most of the time between pulses, hence the average resistance of the switch is high, and the Johnson noise is negligible. In addition, gated detection discriminates effectively against thermal background noise. In fact, van Exter has shown that the thermal background noise usually exceeds the average power of the THz radiation by a factor of ten, and that the minimum detectable THz signal (amplitude) can be 160 times smaller than the incoherent thermal background radiation [2.22].

Because THz-TDS measures electric field rather than intensity, the measurements typically have a greater dynamic range than more conventional techniques. As an example, consider an optically dense medium with a very small power transmission coefficient of $10^{-8}$. A sensitive THz-TDS system can still characterize this medium by measuring the field transmission coefficient of $10^{-4}$.

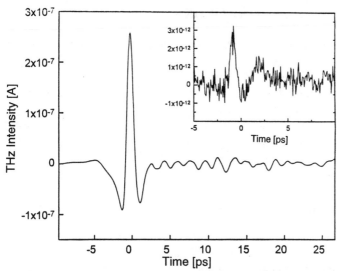

**Fig. 2.10.** Measured THz waveform, with the oscillations in the tail of the waveform resulting from water-vapor absorption in the THz beam path. The scan has nearly $10^6$ : 1 signal-to-noise ratio, as demonstrated in the inset of the figure, obtained by attenuation of the optical beam to a degree that the noise floor becomes visible [2.25]

### 2.3.2 Phase Sensitivity

In many applications, the most important advantage of THz-TDS is direct measurement of the electric field $E(t)$. Fourier transformation of $E(t)$ yields both amplitude and phase of the propagation or transmission coefficients. Measurement of both amplitude and phase in THz-TDS yields real and imaginary parts of the dielectric function over the frequency range spanned by the THz pulse. This is a crucial difference in comparison with conventional FTIR spectroscopy. In FTIR only the *power spectrum* is obtained by Fourier-transformation of the interferogram. Of course, the complex dielectric function can be recovered, in principle, through the use of the Kramers-Kronig relations. However, this requires accurate broad-band measurements, with no way to know, *a priori*, what the frequency range of the measurement must be.

### 2.3.3 Resolution and Time-Windowing of Data

In THz-TDS, the spectral resolution is the inverse of the optical delay time provided by the moving mirror. Because the measurement is performed in the time-domain, substrate reflections can be windowed out of the raw data without much loss in spectral resolution and little influence on the accuracy of the data. This procedure is usually more straightforward than removal of reflections in the frequency-domain.

## 2.4 Time-Domain Data Analysis

Linear spectroscopy requires that the radiation interacts with the medium under study by either reflection or transmission. As with most spectroscopic techniques, THz-TDS requires two measurements: one reference waveform $E_{ref}(t)$ measured without the sample or with a sample of known dielectric properties, and a second measurement $E_{sample}(t)$, in which the radiation interacts with the sample. For spectral analysis, $E(t)$ can directly be Fourier-transformed to yield the complex amplitude spectrum $E(\omega)$ in both amplitude and phase. Depending on the sample, we can distinguish three different cases:

### 2.4.1 Thick Medium

The thick medium is defined as one for which the transit time of the THz pulse is much larger than its duration. Figure 2.11 illustrates the incident, reflected, and transmitted fields at the interfaces. The transmitted radiation consists of a pulse train, with the nth pulse having made $2n+1$ trips through the medium. The Fresnel expressions for reflection and transmission coefficients apply in this situation. Of course, field reflection and transmission coefficients rather than the intensity coefficients must be used. Figure 2.12 shows a typical measurement (after [2.53]). The curve shows the THz transient after propagation through a 283 µm thick silicon wafer. In addition to

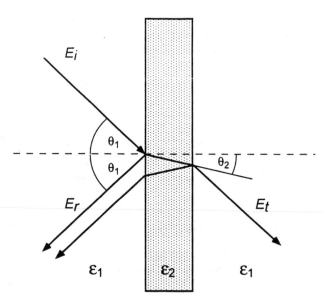

**Fig. 2.11.** Incident, reflected, and transmitted waves at the interface of a thick slab of material. In the thick slab approach, the transit time between front and back interface is much longer than the duration of a THz pulse, so that reflected and transmitted wave from both interfaces do not interfere

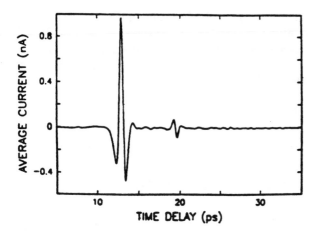

the main transmitted pulse, there is a secondary, time-delayed pulse. This second transient is the first of the infinite series which appears due to multiple reflection. It has traversed the slab twice after being reflected at the back surface. Because the propagation delay is much longer than the pulse duration, the first in the series can be isolated. This pulse has passed through two interfaces and traversed the sample (of thickness $d$) once, emerging with an amplitude given by:

$$E_t = E_0 \cdot t_{12}t_{21} \cdot \exp\left[-i\omega\sqrt{\varepsilon_2(\omega)} \cdot d/c\right],$$ (2.3)

where $E_0$, $E_t$ are incident and transmitted THz fields, $t_{12}$, $t_{21}$ are the complex transmission coefficients at the first and second interface, respectively, $\varepsilon_2$ the complex dielectric constant of medium 2, and $c$ the speed of light. For s-polarization, we find

$$t_{12}^s = \frac{2n_1 \cos\theta_1}{n_1 \cos\theta_1 + n_2 \cos\theta_2}$$ (2.4)

while for p-polarization,

$$t_{12}^p = \frac{2n_1 \cos\theta_1}{n_2 \cos\theta_1 + n_1 \cos\theta_2}.$$ (2.5)

Here, $\theta_1$ and $\theta_2$ is the angle of incidence in medium 1 and 2, respectively.

To obtain the dielectric constant $\varepsilon(\omega)$ of the medium, we must invert (2.3). When absorption and dispersion inside the medium dominates the transmission coefficients at the interfaces, the frequency dependence of the coefficients $t_{ij}$ can often be neglected. In this case, the complex dielectric constant is given by:

$$\sqrt{\varepsilon} = \frac{c}{i\omega d} \cdot \ln\left(\frac{E_t}{E_0 t_{12}t_{21}}\right).$$ (2.6)

More generally, (2.3) needs to be solved iteratively to obtain the dielectric constant $\varepsilon(\omega)$.

For opaque samples it is necessary to analyze the reflected pulse. In this case, the measurement usually compares the THz waveform $E_r(t)$ reflected from the sample, and the waveform $E_{ref}(t)$ reflected from a reference reflector with known reflectivity $r_0(\omega)$ (e.g., a metallic mirror). The complex reflectivity $r(\omega)$ of the sample is then determined from:

$$r(\omega) = \frac{\int\limits_{-\infty}^{\infty} dt\, E_r(t) e^{i\omega t}}{\int\limits_{-\infty}^{\infty} dt\, E_{ref}(t) e^{i\omega t}} \cdot r_0(\omega). \tag{2.7}$$

The Fresnel equations for the reflection coefficients in s, p-polarizations are:

$$r_{12}^s = \frac{n_1 \cos\theta_1 - n_2 \cos\theta_2}{n_1 \cos\theta_1 + n_2 \cos\theta_2}, \tag{2.8}$$

$$r_{12}^p = \frac{n_2 \cos\theta_1 - n_1 \cos\theta_2}{n_2 \cos\theta_1 + n_1 \cos\theta_2}, \tag{2.9}$$

where $n = \sqrt{\varepsilon}$ is the frequency-dependent refractive index. Note that throughout this text, $n$ is complex, including both refractive index and absorption contributions to the dielectric constant. Once the reflection coefficient is determined from (2.7), the Fresnel equations can be inverted to yield the dielectric constant. For example, for an air-dielectric interface and s-polarization, we obtain:

$$\varepsilon(\omega) = \sin^2\theta_1 + \cos^2\theta_1 \left(\frac{1 - r(\omega)}{1 + r(\omega)}\right). \tag{2.10}$$

### 2.4.2 Thin Conducting Sheet

In the limit where the wavelength and the skin depth are much larger than the thickness of the sample, the THz field can be regarded as uniform. Thin conducting films on an insulating substrate are often in this regime. The film separates two media (Fig. 2.13), labeled (1) and (2), in which the fields satisfy the following boundary conditions:

$$\mathbf{n} \times (\mathbf{H}_1 - \mathbf{H}_2) = \int\limits_0^{\infty} J dz \equiv J_S \tag{2.11}$$

$$\mathbf{n} \times (\mathbf{E}_1 - \mathbf{E}_2) = 0 \tag{2.12}$$

where $\mathbf{n}$ is the surface normal vector and $J_S$ is the surface current in the thin film. For s-polarization, we then obtain:

$$(H_i - H_r)\cos\theta_1 - H_t \cos\theta_2 = J_S, \tag{2.13}$$

$$E_i + E_r = E_t,$$

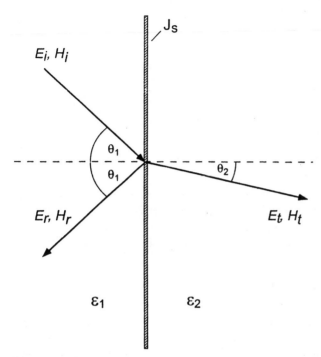

**Fig. 2.13.** Waves at the interface of two materials with a thin conducting sheet at the interface. This approximation is valid for most semiconducting or superconducting films as long as the thickness of the film is much smaller than either the skin depth or the wavelength of the light

and for p-polarization:

$$H_i + H_i - H_t = J_S,$$
$$(E_i - E_r) \cos \theta_1 - E_t \cos \theta_2 = 0. \tag{2.14}$$

The ratio of electric and magnetic fields in the incident, reflected, and transmitted waves is given by the impedance $Z$ (or admittance $Y = 1/Z$) of the two media. To arrive at compact expressions we define the effective admittances $Y_\pm$ as follows:

$$Y_\pm = Y_1 \cos \theta_1 \pm Y_2 \cos \theta_2 , \tag{2.15}$$

where $Y_{1,2}$ is the wave admittance defined by:

$$Y_{1,2} \equiv \sqrt{\frac{\varepsilon_{1,2}}{\mu_{1,2}}}. \tag{2.16}$$

For $\varepsilon = \varepsilon_0$ and $\mu = \mu_0$, this reverts to $Y_0 = (377\,\Omega)^{-1}$, the admittance of free space.

The reflected and transmitted fields can be written in terms of the incident field in the following simple forms (for s-polarization):

$$E_r^s = \frac{1}{Y_+} (Y_- E_i - J_S), \tag{2.17}$$

$$E_t^s = \frac{1}{Y_+} (2Y_1 E_i - J_S), \tag{2.18}$$

and for p-polarization:

$$E_r^p = \frac{1}{Y_1 \cos\theta_2 + Y_2 \cos\theta_1} [(Y_1 \cos\theta_2 - Y_2 \cos\theta_1) E_i - J_S \cos\theta_2], \tag{2.19}$$

$$E_t^p = \frac{\cos\theta_1}{Y_1 \cos\theta_2 + Y_2 \cos\theta_1} (2Y_1 E_i - J_S). \tag{2.20}$$

We can express the surface current in terms of the surface conductivity and the electric field, $J_S = \sigma_S E$, where $\sigma_S$ is the surface conductivity in $\Omega^{-1}$. It is important to recognize that $E$ is the total field at the interface, $E = E_i + E_r$. Then the reflection and transmission coefficients for s-polarization become:

$$r^S = \frac{E_r}{E_i} = \left( \frac{Y_- - \sigma_s}{Y_+ + \sigma_s} \right), \tag{2.21}$$

$$t^S = \frac{E_t}{E_i} = \left( \frac{2Y_1}{Y_+ + \sigma_S} \right). \tag{2.22}$$

Similar expressions for p-polarization can be derived from (2.19, 20). Using $Y = n/377\,\Omega$, the above expressions revert to (2.4–9) when $\sigma_s = 0$.

The above equations can be inverted to obtain the complex surface conductivity of the thin film. We illustrate this procedure using measurements on superconducting niobium films [2.54] as an example. Figure 2.14 shows THz transients transmitted through a niobium film on a sapphire substrate, both above and below the transition temperature $T_c$. In this example, the film in its normal state is used as a reference. Above the superconducting transition the conductivity is frequency-independent and purely real. From (2.22) the ratio of the transmission in the superconducting state (S) and normal state (N) is given by:

$$\frac{t_S}{t_N} = \frac{E_t^S}{E_t^N} = \frac{Y_+ + \sigma_s^N}{Y_+ + \sigma_s^S}. \tag{2.23}$$

The complex conductivity $\sigma(\omega)$ can be obtained directly from the complex Fourier-spectra of the THz waveforms above and below $T_c$ :

$$\frac{\sigma^S(\omega)}{\sigma^N} = \frac{E^S(\omega)}{E^N(\omega)} \cdot \left[ \frac{Y_+}{\sigma_n} \left( 1 - \frac{E^N(\omega)}{E^S(\omega)} \right) + 1 \right]. \tag{2.24}$$

The result of this inversion is shown in Fig. 2.15, showing both real and imaginary part of the conductivity as a function of frequency.

In other experiments, the reference may be the bare substrate, in which case the ratio of the transmission with and without film is:

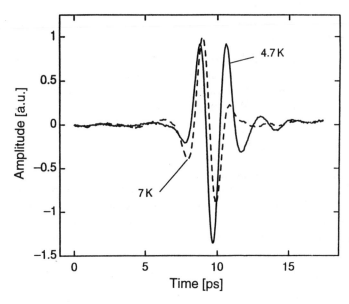

**Fig. 2.14.** Terahertz transients transmitted through a thin niobium film on quartz in the normal state (*dashed*) and in the superconducting state (*solid*) of the super-conductor

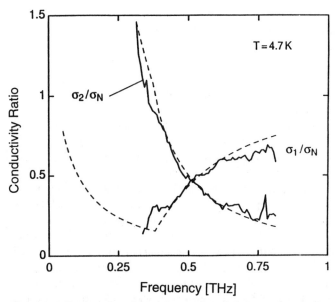

**Fig. 2.15.** Both real and imaginary part of the complex conductivity of the niobium film in the superconducting state, obtained directly from the THz waveforms shown in Fig. 2.14, without use of the Kramers-Kronig relations

$$\frac{E(\omega)}{E'(\omega)} = \frac{Y_+}{Y_+ + \sigma_s} \tag{2.25}$$

with the primed and unprimed field refering to measurements without and with the thin film, respectively. Again, the complex film conductivity can be directly obtained by inverting (2.25). Analogous equations can be found for p-polarization as well as for the reflection geometry.

### 2.4.3 Intermediate Film Thickness

This is the general case of a film of arbitrary thickness on a substrate. We assume that the dielectric constant of the substrate is known. The transmission and reflection coefficients can be derived from the standard theory of propagation through layered media [2.55]. The result is expressed in terms of the Fresnel coefficients, $t_{ij}$ and $r_{ij}$, that describe each interface, and holds for both s- and p-polarizations:

$$r(\omega) = \frac{r_{12} + r_{23}e^{i\Delta}}{1 + r_{12}r_{23}e^{i\Delta}},$$

$$t(\omega) = \frac{t_{12} + t_{23}e^{i\Delta/2}}{1 + r_{12}r_{23}e^{i\Delta}}.$$

Media (1), (2), and (3) refer to vacuum, film, and substrate, respectively. The phase factor $\Delta$ is given by:

$$\Delta = \omega/c \cdot \sqrt{\varepsilon_2} \cos\theta_2 \cdot d,$$

where the $\varepsilon_2$ is the dielectric constant of the film, $\theta_2$ the angle of refraction inside the film, and $d$ its thickness.

This formulation can also describe anisotropic substrates, where the dielectric constant is different for propagation vectors parallel and perpendicular to **n**. An example where this is important is the study of intersubband transitions in quantum wells [2.56]. In this example, the in-plane dielectric constant $\varepsilon_{xy}$ is determined by the Drude conductivity of a free-electron gas, while the dynamics of carriers perpendicular to the plane ($\varepsilon_z$) is determined by quantum confinement. In a medium with anisotropic dielectric constant, cavity modes of the thin slab can interact with resonances of the medium, leading to coupled modes shifted from their position in infinite [2.56, 57]. These coupled modes are referred to as Fuchs-Kliewer modes [2.58].

Unlike the thick medium and thin-film limit, the Fresnel formulae for the general case cannot be inverted for the conductivity or dielectric function. To obtain the dielectric constant of the film in this case, a specific functional form can be assumed for the conductivity and the parameters adjusted numerically for the best fit to the measured THz waveform [2.59].

## 2.5 Linear THz Spectroscopy

THz-TDS offers tremendous advantages relative to other far-infrared spectroscopic techniques in terms of sensitivity, dynamic range, and direct measurement of both real and imaginary parts of the dielectric function. In this section we review examples of successful characterization of the linear response of insulators, semiconductors, superconductors, and other systems using THz-TDS. Most of these measurements were performed using THz beams propagating in free space, although some of the earlier experiments used THz transients in electro-optic crystals.

### 2.5.1 Insulators

Accurate characterization of the far-infrared properties of insulators is needed because of their great technological importance. Insulators with benign properties at terahertz frequencies are crucial to high-frequency applications involving metallic, superconducting, ferromagnetic, and ferroelectric thin films as the active component. Optical phonons, defects, and residual free carriers dominate the processes responsible for the dielectric properties of insulators in this frequency range.

a) **Phonon-Polaritons.** Electromagnetic waves are strongly coupled to the Transverse Optical (TO) phonon modes of ionic insulators. Near the TO frequency phonons and photons mix strongly, leading to new normal modes called "phonon-polaritons." The propagation of phonon-polaritons is observed when THz transients are excited by optical rectification in electro-optic crystals such as $LiTaO_3$ [2.13], as well when free-space THz transients propagate through crystals. By studying the propagation of THz transients, accurate measurements of the polariton dispersion and dephasing can be obtained [2.13].

b) **Optical Materials.** THz-TDS has proved extremely useful in evaluating materials for use as passive optical components in the far-infrared range from 0.1–4 THz. Because of its sensitivity and accuracy, significant deviations from accepted literature values have been discovered. Materials such as sapphire, fused quartz, silicon [2.60], and plastics have been evaluated for transparency and lack of index variation.

As an elemental material, silicon is an ideal optical material because of its weak photon-phonon coupling. The low absorption of silicon in the mid-infrared range is utilized in optical components for infrared imaging and $CO_2$-laser machining at 10 μm. Residual free carriers lead to unacceptable absorption in commercial grade silicon with resistivity less than $1\,000\,\Omega$-cm. THz-TDS showed that floatzone silicon, which can be grown with resistivity in excess of $50\,000\,\Omega$-cm, has excellent optical properties in the far-infrared, with negligible absorption and flat refractive index of 3.42 from 100 GHz to at least 2 THz [2.60]. The large and flat index makes the silicon lenses powerful and free of chromatic aberration.

THz-TDS measurements show that for frequencies below 1–1.5 THz, fused silica optical components are a cheap and simple alternative to high-resistivity silicon [2.60]. Optical alignment with a visible light beam, however, is not possible, because the refractive index in the visible (1.48) and the far-infrared (1.95) are different.

Certain plastics are attractive as optical components in the far-infrared. These materials usually have low reflection losses because of their low dielectric constant and are highly transparent. Another advantage is that they have nearly the same index for refraction in the visible as in the THz region, making visual alignment possible. For example, the polymer TPX (poly-4-methyl-pentene-1) has a low dielectric constant of 1.46 over much of the THz range, with low absorption and dispersion [2.51].

Recently, insulators which are nearly lattice-matched to high-temperature superconducting oxides like $YBa_2Cu_3O_7$ or BaKBiO have received much attention. THz-TDS provides the best broadband characterization of these materials available [2.61]. The most common substrate, $SrTiO_3$, is not suitable for high-frequency applications because of its optically-active soft mode. Alternative substrate materials with lower THz losses include $LaAlO_3$, $NdGaO_3$, MgO, Zirconia, and sapphire [2.60].

A very promising class of materials for high frequency applications, including substrates, are organic polymer films. Some polymers systems are attractive for passive device applications because of their low dielectric constant and ease of processing by spin-coating. In other polymers active devices are possible in which electronic properties are tailored by appropriate doping. The polyimide BPDA-PDA was recently studied in the THz domain using the THz-TDS technique [2.62].

### 2.5.2 Semiconductors

Besides high-resistivity silicon, discussed in Sect. 2.4.1, other semiconductors of interest have been studied by THz-TDS. In particular, the optical properties of Ge and GaAs have been studied up to 2 THz [2.60], and more recently, the absorption of semi-insulating GaAs up to a frequency of 5 THz has been investigated [2.63]. The THz frequency range is particularly interesting for doped semiconductors, as the majority of the oscillator strength of the Drude conductivity falls in this frequency range. The carrier concentration and scattering rates can be determined from the frequency-dependent conductivity, $\sigma(\omega)$, in this range. As a result, THz-TDS is well suited to noncontact doping and mobility measurements of semiconductor wafers.

THz-TDS, because of its high accuracy, allows a careful examination of possible deviations of $\sigma(\omega)$ from the Drude formula. Such deviations can point to energy-dependent scattering rates [2.53, 64], or non-equilibrium distribution functions which differ from the predictions of Boltzmann transport theory. At room temperature, it was shown that the conductivity of doped GaAs follows the Drude formula closely even up to 4 THz [2.65]. However, deviations are expected at low temperatures, in particular in modulation-doped semiconductors. While THz-TDS measurements at low temperature have been performed in modulation-doped structures [2.56, 66–68], to date they have not focused on the frequency dependence of the conductivity of the electron gas.

### 2.5.3 Quantized Levels in Confined Systems

Semiconductor nanostructures have attracted interest for many years due to the physics involved in quantum confinement and applications like the Quantum-Confined Stark Effect (QCSE) [2.69]. Structures with confinement in two (quantum wires) and three dimensions (quantum dots) have recently become available. Because the quantum confinement energy is typically in the THz range, THz-TDS is an ideal tool for the study of semiconductor nanostructures.

The first THz-TDS experiment in a quantum-confined system measured transitions between the levels of a coupled quantum-well structure [2.56]. The far-infrared response of such a system can be quite complicated due to the interplay of free-electron motion parallel to the wells and confinement perpendicular to the wells. The dielectric function is influenced by many-body interactions [2.57, 70], and by the interaction with cavity (Fuchs-Kliewer) modes of the thin semiconductor film. These interactions shift the absorption away from where the single-particle transitions are expected to occur. More recently, the dynamical conductivity of a two-dimensional, high-mobility electron gas has been measured in the presence of a magnetic field [2.66]. In these experiments, transitions between Landau levels were observed, and the time-dependent longitudinal and transverse conductivities $\sigma_{xx}(t)$ and $\sigma_{xy}(t)$ were measured.

### 2.5.4 Photonic Band Gaps

A new type of artificial material with interesting properties are the periodic dielectric structures known as "photonic crystals." These materials exhibit bands in which propagation of electromagnetic waves is forbidden, at least in certain directions. Structures with photonic bandgaps have several potential applications, for example the inhibition of spontaneous emission in semiconductor lasers [2.71]. Photonic crystals with bandgaps in the visible range have unit cells on the micron length scale and fabrication has proved difficult. The concept has been tested, however, in periodic structures with lattice parameters in the 0.1–1 mm range, where fabrication is simpler. In these systems the bandgaps lie in the GHz to THz regime. Two-dimensional photonic crystals made from alumina rods were investigated with optoelectronic TDS optimized at a few tens of GHz [2.72–75]. Using advanced microfabrication techniques, Ozbay has built three-dimensional photonic crystals in silicon and demonstrated photonic bandgaps in these structures using all-electronic THz-TDS [2.76, 77].

### 2.5.5 THz Spectroscopy of Superconductors

The optical properties of superconductors have been studied since their discovery by Kammerling-Onnes in 1911. Far-infrared spectroscopy [2.78, 79] was instrumental in the discovery of the superconducting energy gap. In these pioneering experiments, significant extrapolation was required in order to extract the real and imaginary parts of the conductivity $via$ the Kramers-Kronig relations. Using THz-TDS, the complex $\sigma(\omega)$ in the superconducting state is determined with almost astonishing simplicity and accuracy, as was demonstrated in niobium [2.54]. As seen in Fig. 2.15, the onset of dissipation $(\mathrm{Re}\{\sigma\})$ at the gap edge is clearly observed despite the fact that the response function is dominated by inductive screening $(\mathrm{Im}\{\sigma\})$.

With the discovery of high-$T_c$ oxide superconductors, far-infrared spectroscopy became prominent again in the study of superconductivity. THz-TDS has played a central role in uncovering the complex properties of these materials. The emphasis of THz-TDS has been different than the previous generation of far-infrared experiments, which focused on measuring the onset of $\mathrm{Re}\{\sigma\}$ that defines the gap. In high-$T_c$ materials the gap is expected to lie in the range between 10 and 15 THz, above the high-frequency limit of THz-TDS. However, THz-TDS is ideally suited to study processes involving "quasi-particles," or thermally-dissociated Cooper pairs. The quasiparticle scattering rate is now known to vary from about 100 GHz at low temperatures to 3 THz near $T_c$ in the 90 K superconductor $YBa_2Cu_3O_7$ (YBCO), almost a perfect match to the range of THz-TDS.

One of the most influential measurements in the study of cuprate superconductivity was the THz-TDS measurement of Re and Im$\{\sigma\}$ from 10–120 K in YBCO [2.80]. The measurements were performed in the transmission mode, on high-quality films in the thin-conductor limit (500 Å). It was shown that $\mathrm{Re}\{\sigma\}$, upon cooling through $T_c$, rose to a peak near 60 K and then fell as the temperature was lowered further. Although this behavior superficially resembles the "coherence peak" predicted by Mattis and Bardeen for "dirty-limit" superconductors, it was soon appreciated that its origin was completely different. In fact, the peak arises from a rapid drop in the quasiparticle scattering rate which takes place near $T_c$ [2.80–82]. This observation demonstrated that quasiparticle scattering in high-$T_c$ cuprate superconductors is dominated by electron-electron, rather than electron-phonon scattering.

Because of its power to measure quasiparticle dissipation in the cuprates, THz-TDS continues to play an important role in the study of high-$T_c$ superconductors. Recent experiments have focused on new materials and the effects of impurity doping and magnetic fields. THz-TDS measurements in BaKBiO

[2.83] are intriguing in that, like the cuprates, a large enhancement in the real part of the conductivity $(\sigma_1)$ is seen below $T_c$. In another important application, the terahertz conductivity has been investigated in "underdoped" cuprates, i.e., samples whose carrier concentration is below that required to optimize $T_c$ [2.84]. In this experiment the hole concentration in YBCO is modified by doping with Pr. Evidence is found for an increase in conductivity above $T_c$, which suggests that the "spin-gap" plays a role in the charge fluctuation spectrum.

Furthermore, THz-TDS is ideally suited to the study of superconductors in a magnetic field. Because the field breaks time-reversal symmetry, the electrodynamic response is characterized by two complex response functions, $\sigma_{xx}$ and $\sigma_{xy}$ (or $\sigma_\pm$), rather than one. Using THz-TDS based on free-space propagation, both functions can be determined directly from polarization-resolved transmission measurements.

Figure 2.16 shows a schematic of the system used to measure both the diagonal and off-diagonal components of the transmission tensor [2.85]. To measure $t_{xy}$ and $t_{xx}$, the incident polarization is aligned perpendicular and parallel to the analyzer, respectively. Detecting electric-field amplitude rather than intensity is crucial to a sensitive measurement of $\sigma_{xy}$. The field transmission coefficient, $t_{xy}$ is proportional to the Faraday rotation angle, $\theta_F \approx \sigma_{xy}/\sigma_{xx}$, for small angles. If intensity rather than field is detected, the signal is proportional to $(\sigma_{xy}/\sigma_{xx})^2$ and small rotations are much more difficult to detect. Although linearity in $\theta_F$ is recovered if the incident field is biased at $\pi/4$ relative to the analyzer, the Faraday rotation signal is now superposed on a large background of transmission. Recent studies using polarization-resolved transmission spectroscopy have detected the cyclotron motion of quasiparticles [2.85] and probed the high-frequency dynamics of vortices [2.86].

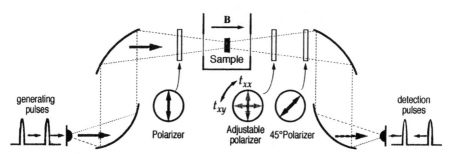

**Fig. 2.16.** Setup used for Faraday rotation spectroscopy [2.85]. This setup can measure the transmission tensor $t_{xy}$ of samples in a magnetic field

## 2.6 Time-Resolved Optical Pump – THz Probe Spectroscopy

The measurements discussed so far show that THz-TDS can, in many cases, improve upon standard far-infrared spectroscopy by offering larger S/N-ratios and direct measurement of the complex response tensor. In contrast, the optical pump – THz probe measurements described in this section are simply impossible with conventional techniques, as they rely on the time-domain and optoelectronic nature of THz-TDS. Time-resolved THz pump-probe experiments hold great promise for the investigation of time-dependent phenomena which cannot be accessed with standard far-infrared techniques [2.12, 87].

An example is the measurement of the frequency-dependent conductivity of photocarriers as a function of time after their generation. A setup for this experiment, which is a straightforward extension of the linear THz-TDS spectrometer described previously, is shown in Fig. 2.17. A beam from the same laser that triggers the THz transmitter and detector photoexcites the sample. Using delay lines, the frequency-resolved conductivity can be obtained at any time after the optical excitation, with femtosecond resolution. In a more elaborate setup, the frequency of the excitation beam may be different from the gating beams. This can be accomplished by pumping an optical para-

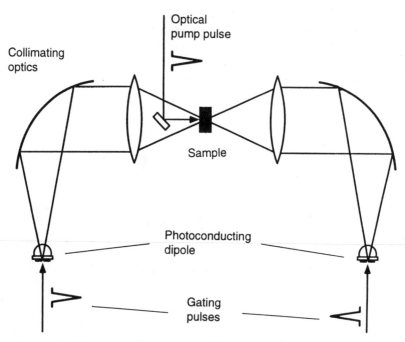

**Fig. 2.17.** Schematic diagram of a THz pump-probe experiment. An additional optical pump pulse directly excites the sample, and the focused THz beam probes the far-infrared properties of the sample as a function of time after optical excitation

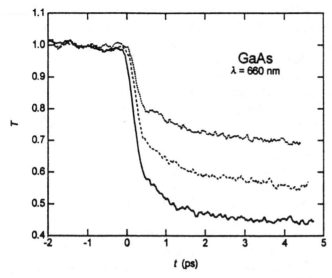

**Fig. 2.18.** Time-dependent carrier mobility in GaAs after injection of hot carriers into the conduction band. The 2 ps time constant can be identified with the transfer of electrons form the low-mobility $L$ and $X$ valleys to the high-mobility $\Gamma$-valley [2.87]

metric oscillator with part of the original beam, or by some other frequency conversion process.

Using a similar optical-THz pump-probe technique, the time and frequency-dependent conductivity of hot electrons in GaAs and InP was studied [2.12, 2.87–89]. These measurements showed that when the excess energy was large enough, electrons were transferred into satellite valleys within less than 100 fs. These low-mobility electrons took about 1–2 ps to return to the central $\Gamma$-valley (Fig. 2.18). The fraction of the electrons being transferred was found to vary with the excess energy, and the return times were found to be influenced by electron cooling rates in the $\Gamma$-valley [2.87].

Another interesting series of pump-probe experiments was performed in superconducting lead films. In this case, the optical pump leads to photo-induced dissociation of Cooper pairs and time-resolved THz-TDS measures the return to the equilibrium state [2.87]. As a final example, Groeneveld et al. time-resolved the far-infrared transition between 1 s and 2 p states of optically generated excitons in GaAs-AlGaAs quantum wells, and deduced exciton creation and decay times [2.90].

## 2.7 THz Correlation Spectroscopy

Another type of measurement, THz correlation spectroscopy (THz-CS), is a hybrid between THz-TDS and FTIR spectroscopy. This technique is conceptually the same as FTIR in that two beams, with a variable time-delay between them, are combined with a beam-splitter and imaged onto a "square-law" or power detector. In FTIR the two beams are generated by splitting the beam from a single thermal source and the time delay generated by varying the path length of one of the beams. In THz-CS, Auston switch sources replace the thermal source. In both cases the detector current as a function of path length is called an "interferogram." The Fourier transform of the interferogram gives the product of the spectral density of the source and the spectrometer throughput function, including the response of the mirrors, detector, and other optical elements.

The goal of the first experiment to use this technique [2.28] was to resolve, for the first time, the true spectral density of a THz-pulse generator. As we have mentioned, the full bandwidth of THz pulsed sources cannot be measured by gated-optoelectronic detection because of the nonzero photo-carrier lifetime. To find the spectral density of optoelectronic sources, *Greene* et al. substituted them for the thermal source in a conventional Michelson FTIR spectrometer. The recombined beam from the interferometer was imaged onto a $^4$He-cooled bolometer which produces a voltage proportional to the average incident power. The resolution of the measurement is limited only by the path length of the scan. This experiment was able to verify that radiation emitted from an unbiased InP wafer optoelectronic source extends to frequencies as high as 7 THz [2.28]. More recently, this technique has been utilized to show that the spectral bandwidth of a similar source extends even further into the mid-IR range, when optical pulses as short as 15 fs are used for the optical excitation [2.41].

While the goal of these experiments is to characterize optoelectronic sources, THz-CS is ideally suited as well to photocurrent spectroscopy. A photocurrent spectrum is the change in the dc conductance of a structure or device due to photoexcitation, measured as a function of the photon frequency. A typical set-up to measure the photocurrent spectrum using THz-CS is shown in Fig. 2.19 [2.91]. A device under study is irradiated by a pair of THz pulses, originating in two Auston switch sources whose beams are combined using a beam splitter. The advantage of this system compared to the Michel-

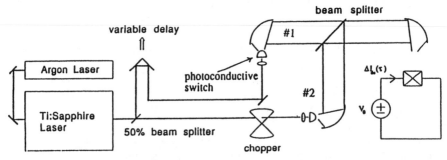

**Fig. 2.19.** THz-CS setup [2.91]

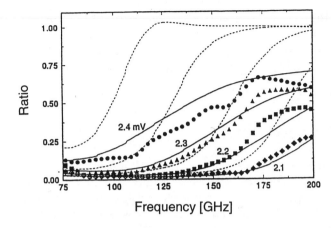

**Fig. 2.20.** Nb-NbO-Nb THz-CS photocurrent response [2.92]

son is that the variable time-delay can be accomplished in the laser, rather than infrared, beam-path. The inset in Fig. 2.19 shows the current generated by irradiation of the device with the pulse pair, as a function of the time delay between them. The Fourier transform of this trace is the photocurrent spectrum of the device.

Photocurrent spectroscopy is particularly useful in the study of "mesoscopic" systems – materials or structures with typical dimensions of 10–1000 nm. Many of these systems possess energy gaps or collective modes in the meV range, making them ideal candidates for far-infrared spectroscopy. However, mesoscopic structures are far smaller than the wavelength of the radiation, making them entirely unsuitable for transmission or reflection spectroscopy. By contrast, the fractional conductance change of the structure due to photoexcitation is independent of the sample size.

To date, two types of mesoscopic structures have been studied using THz-CS, Superconductor-Insulator-Superconductor (SIS) junctions, [2.92] and Quantum Point Contacts (QPC's). [2.91, 93] The former represent the first broadband measurements of the response of an SIS detector. In these devices the Josephson current is turned off through the application of a small magnetic field. The photocurrent results from photon-assisted tunneling of quasi-particles from one side of the junction to the other. The threshold frequency for observing the photocurrent is $2\Delta\, eV$, where $2\Delta$ is the energy gap, $e$ the electron charge, and $V$ the applied voltage. The spectra in Fig. 2.20 show the photocurrent response of a Nb-NbO-Nb junction [2.92]. The fits were obtained using the theory for photon-assisted tunneling [2.94] with essentially no free parameters. All of the parameters of the theory are obtained from the dc current-voltage characteristic of the junction.

Photocurrent spectra have also been obtained in a quantum point contact using THz-CS [2.91, 93]. The QPC is a narrow constriction separating two regions of electron gas. In this case the QPC was formed by defining narrow metallic gates above the Two-Dimensional Electron Gas (2DEG) in a GaAs/AlGaAs heterojunction. Evidence for photon-assisted tunneling was found from the frequency shift in the photocurrent spectra as a function of voltage applied to the gate. Modulation of the photocurrent was also observed when the applied radiation was resonant with magnetoplasmon modes of the 2DEG.

## 2.8 THz Emission Spectroscopy

THz absorption in a medium implies the existence of allowed dipole transitions. Conversely, if we excite a time-varying far-infrared dipole moment, this time-varying dipole will emit radiation at the oscillation frequency of the dipole. The principle of THz emission spectroscopy is that such a time-varying dipole moment can be excited using an optical pulse. A very simple example is the previously discussed optical generation of carriers that are subsequently accelerated in an applied electric field, hence leading to a time-varying dipole moment $P(t)$. This time-varying dipole moment leads to a radiated field $E(t) \propto d^2 P/dt^2$. A more complicated yet interesting situation arises when a short optical pulse excites a manifold of quantum states in the medium. The coherent excitation of these states leads to "quantum beats" of all the involved optical transitions [2.95]. The beat frequencies may lie in the THz range, but THz radiation is only emitted when there are allowed far-infrared dipole transitions within the manifold.

THz emission following excitation with a short laser pulse can also be viewed as a nonlinear difference-frequency process in which two (typically visible) photons generate a THz photon. The process is often referred to as "optical rectification." The usual geometry used for THz emission spectroscopy is the large-aperture antenna (Fig. 2.5). Frequently, the dipole moments in these experiments are perpendicular to the surface, and emission of dipole radiation can only be observed away from the surface normal because of the $\sin^2 \theta$ angular dependence of the dipole radiation pattern.

In general, the THz emission spectrum contains information similar to the absorption spectrum. If the dipole moment of the transition of interest is small and obscured by a large background, THz emission usually gives more reliable results than absorption spectroscopy. In some cases, the nonlinear generation process can also reveal a wealth of information which cannot be obtained from linear spectroscopy, regarding interactions of carriers and excitons [2.96].

Two examples of THz Emission Spectroscopy have already been discussed: the generation of THz radiation from acceleration of photocarriers in an applied electric field [2.30, 31, 33, 35, 97, 98], from phonon-polaritons in LiTaO$_3$ [2.13, 99–101], and from optical phonons in Tellurium [2.102]. Other examples of THz emission spectroscopy are THz transients generated by optical excitation of organic crystals [2.103], and emission of THz radiation from charge oscillations in quantum-well structures [2.37, 104–111].

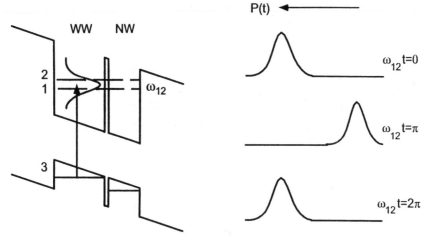

**Fig. 2.21.** Coupled Quantum Well (CQW) structure [2.104]. After excitation of excitons in the Wide Well (WW), the excitonic wavepacket oscillates back and forth between both wells at the splitting frequency. The resulting oscillating dipole moment leads to emission of THz radiation, centered at the splitting frequency

The emission of THz radiation from charge oscillations in quantum-well structures is particularly interesting, and serves to illustrate the principle of this technique. One of the structures investigated is the Coupled Quantum-Well (CQW), in which two GaAs quantum-wells of different width are separated by a thin AlGaAs barrier (Fig. 2.21) [2.104]. A dc electric field can be applied to the structure to bring electronic states in the narrow well and the wide well into resonance. The wavefunctions now interact strongly, leading to a splitting of the states into symmetric and antisymmetric combination states.

Because the interband transition energy is lower in the Wide Well (WW) than the Narrow Well (NW), excitons can be selectively injected into the wide well. If the femtosecond laser spectrum is wider than the energy splitting between the symmetric and antisymmetric wavefunctions of the coupled well, a superposition of these two wavefunctions is excited, which evolves in time. After photoexcitation, the electron component of the exciton oscillates back and forth between the two wells at the splitting frequency, while the hole remains localized in the WW. This motion constitutes a time-varying dipole moment, which will radiate at the splitting frequency of the CQW. Figure 2.22 shows an example of the THz waveforms [2.104] emitted from a CQW following excitation by a 100 fs laser pulse centered at $\lambda = 810$ nm (the interband transition wavelength for the WW). More than ten cycles of the radiation can be observed before dephasing, each corresponding to an oscillation of the electron between the wells.

The results of this and other experiments can be understood by a simple three-level model (Fig. 2.23), in which the lower level $|3\rangle$ represents the local-

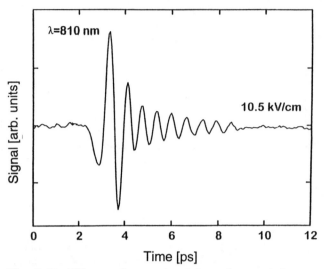

**Fig. 2.22.** THz waveform emitted from the coupled quantum well structure. As many as 10 oscillations can be observed before the excitonic wave packet looses its phase coherence [2.104]

ized hole state in the valence band, and the two upper states $|1\rangle$ and $|2\rangle$ are the symmetric and antisymmetric eigenstates of electrons in the conduction band. Because the width in frequency of the femtosecond laser pulse spans the two excited states, optical excitation generates a coherent superposition of $|1\rangle$ and $|2\rangle$. The subsequent time-evolution of the system shows "quantum beating" phenomena at the splitting frequency. If the far-infrared dipole moment $\mu_{12}$ is non-vanishing, the system radiates at the splitting frequency. Excitonic effects change the splitting levels and the effective splitting fields significantly when compared to the single-particle picture [2.104]. However, the fact that excitons rather than free electron-hole pairs are excited is responsible for the relatively long (picosecond) dephasing times of the charge oscillations observed in these experiments.

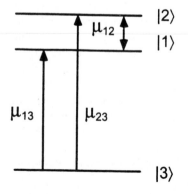

**Fig. 2.23.** Three-level system illustrating the various transitions involved in THz emission spectroscopy. The dipole moment $\mu_{12}$ has to be nonzero in order to observe oscillatory THz emission

THz emission from quantum-confined semiconductor nanostructures has also been seen due to quantum beats between heavy and light holes states [2.105, 108], and from charge oscillations in semiconductor superlattices [2.110, 111]. Interesting results have also been obtained recently from coherently excited Landau-levels of semiconductor quantum wells in a magnetic field [2.67, 68].

## 2.9 "T-Ray" Imaging

The high signal-to-noise ratio and sensitivity of THz-TDS permits extremely rapid data acquisition. *Hu* et al. [2.112, 113] have taken advantage of these high acquisition and processing rates to build a system which performs two-dimensional imaging. In this system, it is possible to obtain between 10 and 100 spectra per second with a S/N ratio of greater than 1000:1. The data handling is based on a Digital Signal Processor (DSP) card available for personal computers which perform acquisition and A/D conversion at a rate of > 50 kSamples/s, without the use of a lock-in analyzer. Fast Fourier Transforms (FFT) or other analysis of the acquired waveforms are accomplished in real time using the on-board DSP.

Figure 2.24 shows a schematic of the setup, which is similar to the terahertz spectrometer shown in Fig. 2.1. An important addition is the scanning delay line, which rapidly scans the optical delay between the pulses triggering the THz transmitter and receiver, respectively. This effectively downconverts the THz waveforms into the kHz (audio) range, which allows acquisition and processing of up to 100 terahertz waveforms per second with a signal-to-noise ratio exceeding 1000:1. To maximize the spatial resolution, 12 cm focal length paraboloids are used to collect the THz radiation and f = 6.6 cm paraboloids

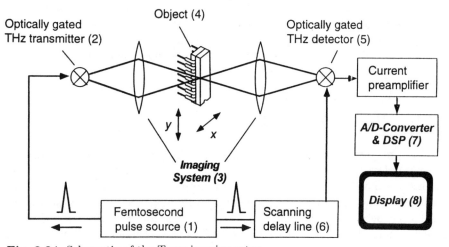

**Fig. 2.24.** Schematic of the T-ray imaging setup

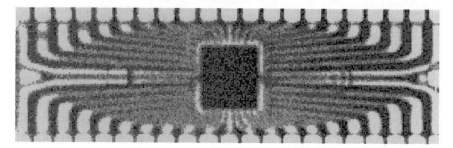

**Fig. 2.25.** T-ray image of a packaged semiconductor chip. Clearly, the silicon chip, as well as the metal leads can be seen through the plastic package

are used to focus the radiation onto the sample. The spatial resolution is roughly 300 μm, close to the diffraction limit for 1 THz radiation. The sample is scanned in $x$ and $y$ by a motor-driven translation stage at a rate of up to 100 pixels/second, so that a 100 × 100 pixel image takes just over a minute to acquire.

Figure 2.25 shows a terahertz "T-ray" image of a plastic-packaged semiconductor chip obtained by this technique. The plastic packaging material is largely transparent to the THz pulses, while the silicon chip and the metal leads are opaque and are clearly revealed within the package.

The power of the T-ray technique is not only in the extension of THz spectroscopy to imaging, but in the real-time processing that is available using the DSP. Because each material has a distinct absorption and dispersion curve over the 100 GHz to 4 THz range sampled by a pulse, each THz pulse also has a characteristic shape after passing through the material [2.113]. Thus, "recognizing" a material reduces to a recognition of the shape of the THz pulse after traversing the sample. Of course, this will not always be a unique identification, in particular in the case of mixtures. Interestingly enough, the algorithms used for speech recognition are well suited for recognizing THz waveforms as well, and much of the software research necessary to make T-ray chemical imaging a reality already exists.

In the future, it may also be possible to acquire the entire T-ray image simultaneously using detector arrays.

## 2.10 All-Electronic THz-TDS

Recent technological advances have enabled the generation of ultrashort electrical pulses without the use of femtosecond lasers. These all-electronic systems are based on the physics of pulse-propagation in nonlinear media. In this case, the medium is a nonlinear transmission line. The nonlinearity is achieved by periodically shunting a conventional line with nonlinear devices such as ultra-high frequency Schottky or Resonant Tunneling Diodes (RTDs)

[2.114]. A sinusoidial wave, generated by a high frequency synthesizer, will undergo compression in the time-domain due to the nonlinearity. The compression is somewhat similar to soliton formation in optical fibers, arising from the nonlinear refractive index of the fiber material.

Because the technique is new, there are only few reports of materials characterization using this technique [2.76, 115, 116]. It is probably too early to evaluate the potential of this technique to replace laser-based systems. There appear to be some disadvantages: it may be difficult to generate significant power beyond 2 THz, and the ultimate signal-to-noise ratio and reproducibility are not known. The big advantage, of course, is the elimination of the femtosecond laser. However, with the current progress and commercialization of compact, diode-pumped femtosecond optical sources, this advantage may evaporate quickly. One hopes that progress in both areas will ultimately result in affordable and user-friendly THz spectrometers on the market in the near future.

## 2.11 Outlook and Summary

Since the invention of THz time-domain spectroscopy in the late 1980s, THz-TDS has made significant contributions to materials and device characterization in the far-infrared range of the electro-magnetic spectrum. This development is intimately tied to the proliferation and commercialization of femtosecond laser sources. This trend is expected to continue with the development of compact, diode-pumped femtosecond sources that will make THz-TDS easier to use, more affordable, and more practical.

In addition to these technological developments, optoelectronic THz techniques have fostered a number of new and innovative techniques like pump-probe spectroscopy, THz emission spectroscopy, and THz coherence spectroscopy. This trend is likely to continue, but we also foresee that THz-TDS will have real-world applications outside of the scientific community, as can be seen from the example of terahertz "T-ray" imaging.

## References

2.1   D. H. Auston: Ultrafast optoelectronics, in *Ultrashort Laser Pulses and Applications*, 2nd edn., ed. by W. Kaiser, Topics Appl. Phys., Vol.60 (Springer, Berlin, Heidelberg 1993)

2.2   D. H. Auston: Appl. Phys. Lett. **26**, 101-103 (1975)

2.3   G. Mourou, C. V. Stancampiano, A. Antonetti, A. Orszag: Picosecond microwave pulses generated with a subpicosecond laser-driven semiconductor switch, Appl. Phys. Lett. **39**, 295-296 (1981)

2.4   R. Heidemann, T. Pfeiffer, D. Jäger: Optoelectronically pulsed slot-line antennas, Electron. Lett. **19**, 316-317 (1983)

2.5    A.P. DeFonzo, M. Jarwala, C. Lutz: Transient response of planar integrated opto-electronic antennas, Appl. Phys. Lett. **50**, 1155-1157 (1987)

2.6    Y. Pastol, G. Arjavalingam, J.M. Halbout, G.V. Kopcsay: Characterization of an optoelectronically pulsed broadband microwave antenna, Electron. Lett. **24**, 1318-1319 (1988)

2.7    D.H. Auston: Picosecond photoconducting Hertzian dipoles, Appl. Phys. Lett. **45**, 284-286 (1984)

2.8    P.R. Smith, D.H. Auston, M.C. Nuss: Subpicosecond photoconducting dipole antennas, IEEE J. QE-**24**, 255-260 (1988)

2.9    C. Fattinger, D. Grischkowsky: Point source terahertz optics, Appl. Phys. Lett. **53**, 1480-1482 (1988)

2.10   C. Fattinger, D. Grischkowsky: Terahertz beams, Appl. Phys. Lett. **54**, 490-492 (1989)

2.11   D.H. Auston, K.P. Cheung, J.A. Valdmanis, D.A. Kleinman: Cherenkov radiation from femtosecond optical pulses in electro-optic media, Phys. Rev. Lett. **53**, 1555-1558 (1984)

2.12   M.C. Nuss, D.H. Auston, F. Capasso: Direct subpicosecond measurement of carrier mobility of photoexcited electrons in gallium arsenide, Phys. Rev. Lett. **58**, 2355-2358 (1987)

2.13   D.H. Auston, M.C. Nuss: Electro-optical generation and detection of femtosecond electrical transients, IEEE J. QE-**24**, 184-197 (1988)

2.14   R.L. Fork, C.V. Shank, B.I. Greene: Appl. Phys. Lett. **38**, 671-672 (1981)

2.15   D.E. Spence, P.N. Kean, W. Sibbett: 60 fsec pulse generation from a self-mode-locked Ti:Sapphire laser, Opt. Lett. **16**, 42-44 (1991)

2.16   W.H. Knox: The revolution in femtosecond near-infrared pulse generation, Opt. Photon. News **3**, 10-14 (May 1992)

2.17   D. Kopf, K.J. Weingarten, L.R. Brovelli, M. Kamp, U. Keller: Diode-pumped 100-fs passively mode-locked Cr:LiSAF laser with an antiresonant Fabry-Perot saturable absorber, Opt. Lett. **19**, 2143-2145 (1994)

2.18   N. H. Rizvi, P. M. W. French, J. R. Taylor: 50-fs pulse generation from a self-staring cw passively mode-locked Cr:LiSrAlF$_6$ laser, Opt. Lett. **17**, 877-879 (1992)

2.19   S. Tsuda, W.H. Knox, E.A. De Souza, W.Y. Jan, J.E. Cunningham: Low-loss intracavity AlAs/AlGaAs saturable Bragg reflector for femtosecond mode locking in solid-state lasers, Opt. Lett. **20**, 1407-1409 (1995)

2.20   D.H. Auston: Picosecond photoconductors: Physical properties and applications, in *Picosecond Optoelectronic Devices*, ed. by C.H. Lee (Academic, Orlando, FL 1984) pp.73-117

2.21   D.H. Auston: IEEE J. QE-**19**, 639-648 (1983)

2.22   M. van Exter, D.R. Grischkowsky: Characterization of an optoelectronic terahertz beam system, IEEE Trans. MTT-**38**, 1684-1691 (1990)

2.23   N. Katzenellenbogen, D. Grischkowsky: Efficient generation of 380 fs pulses of THz radiation by ultrafast laser pulse excitation of a biased metal-semiconductor interface, Appl. Phys. Lett. **58**, 222-224 (1991)

2.24   S.E. Ralph, D. Grischkowsky: Trap-enhanced electric fields in semi-insulators: the role of electrical and optical carrier injection, Appl. Phys. Lett. **59**, 1972-1974 (1991)

2.25    I. Brener, D. Dykaar, A. Frommer, L.N. Pfeiffer, J. Lopata, J. Wynn, K. West, M. C. Nuss: Terahertz emission from electric field singularities in biased semiconductors, Opt. Lett. **21**, 1924-1926 (1996)

2.26    D. Krökel, D. Grischkowsky, M.B. Ketchen: Subpicosecond electrical pulse generation using photoconductive switches with long carrier lifetimes, Appl. Phys Lett. **54**, 1046-1047 (1989)

2.27    B.B. Hu, J.T. Darrow, X.C. Zhang, D.H. Auston: Optically steerable photoconducting antennas, Appl. Phys. Lett. **56**, 886-888 (1990)

2.28    B.I. Greene, J.F. Federici, D.R. Dykaar, R.R. Jones, P.H. Bucksbaum: Interferometric characterization of 160 fs far-infrared light pulses, Appl. Phys. Lett. **59**, 893-895 (1991)

2.29    D. You, R.R. Jones, P.H. Bucksbaum, D.R. Dykaar: Generation of high-power sub-single-cycle 500-fs electromagnetic pulses, Optics Lett. **18**, 290-292 (1993)

2.30    X.C. Zhang, B.B. Hu, J.T. Darrow, D.H. Auston: Generation of femtosecond electromagnetic pulses from semiconductor surfaces, Appl. Phys. Lett. **56**, 1011-1013 (1990)

2.31    J.T. Darrow, B.B. Hu, X.C. Zhang, D.H. Auston: Subpicosecond electromagnetic pulses from large-aperture photoconducting antennas, Optics Lett. **15**, 323-325 (1990)

2.32    L. Xu, X. C. Zhang, D.H. Auston: Terahertz radiation from large aperture Si p-i-n diodes, Appl. Phys. Lett. **59**, 3357-3359 (1991)

2.33    X.C. Zhang, J.T. Darrow, B.B. Hu, D.H. Auston, M.T. Schmidt, P. Tham, E.S. Yang: Optically induced electromagnetic radiation from semiconductor surfaces, Appl. Phys. Lett. **56**, 2228-2230 (1990)

2.34    R.R. Jones, N.E. Tielking, D. You, C. Raman, P. H. Bucksbaum: Ionization of oriented Rydberg states by subpicosecond half-cycle electromagnetic pulses, Phys. Rev. A **51**, R2687 (1995)

2.35    B.B. Hu, E.A. de Souza, W.H. Knox, J.E. Cunningham, M.C. Nuss, A.V. Kuznetsov, S.L. Chuang: Identifying the distinct phases of carrier transport in semiconductors with 10 fs resolution, Phys. Rev. Lett. **74**, 1689-1692 (1995)

2.36    D. Grischkowsky, N. Katzenellenbogen: Femtosecond pulses of terahertz radiation: physics and applications, in *Picosecond Electronics and Optoelectronics IX* (Salt Lake City, UT (1991) (OSA, Washington, DC 1991) pp.9-14

2.37    P.C.M. Planken, M.C. Nuss, W.H. Knox, D.A.B. Miller, K.W. Goossen: THz pulses from the creation of polarized electron-hole pairs in biased quantum wells, Appl. Phys. Lett. **61**, 2009-2011 (1992)

2.38    S.L. Chuang, S. Schmitt-Rink, B.I. Greene, P.N. Saeta, A.F.J. Levi: Optical rectification at semiconductor surfaces, Phys. Rev. Lett. **68**, 102-105 (1992)

2.39    B.B. Hu, A.S. Weling, D.H. Auston, A. V. Kuznetsov, C.J. Stanton: DC-electric-field dependence of THz radiation induced by femtosecond optical excitation of bulk GaAs, Phys. Rev. B **49**, 2234-2237 (1994)

2.40    B.B. Hu, X.C. Zhang, D.H. Auston: Terahertz radiation induced by sub band-gap femtosecond optical excitation of GaAs, Phys. Rev. Lett. **67**, 2709-2711 (1991)

2.41    A. Bonvalet, M. Joffre, J.L. Martin, A. Migus: Generation of ultrabroadband femtosecond pulses in the mid-infrared by optical rectification of 15 fs light pulses at 100 MHz repetition rate, Appl. Phys. Lett. **67**, 2907-2909 (1995)

2.42   S. Gupta, M.Y. Frankel, J.A. Valdmanis, J.F. Whitaker, G.A. Mourou, F.W. Smith, A.R. Calawa: Subpicosecond carrier lifetime in GaAs grown by molecular beam epitaxy at low temperatures, Appl. Phys. Lett. **59**, 3276-3278 (1991)

2.43   F.E. Doany, D. Grischkowsky, C.C. Chi: Carrier lifetime versus ion-implantation dose in silicon on sapphire, Appl. Phys. Lett. **50**, 460-462 (1987)

2.44   D.R. Dykaar, B.I. Greene, J.F. Federici, A.F.J. Levi, L.N. Pfeiffer, R.F. Kopf: Log-periodic antennas for pulsed terahertz radiation, Appl. Phys. Lett. **59**, 262-264 (1991)

2.45   D.B. Rutledge, D.P. Neikirk, D.P. Kasilingam: Integrated-Circuit Antennas, in *Infrared and Millimeter Waves* **10** (Academic, Orlando, FL 1983)

2.46   C. Fattinger, D. Grischkowsky: Beams of terahertz electromagnetic pulses, in *Picosecond Electronics and Optoelectronics IV* (Salt Lake City, UT, USA (1989) (OSA, Washington, DC 1989) pp.225-231

2.47   P.U. Jepsen, S.R. Keiding: Radiation patterns from lens-coupled terahertz antennas, Opt. Lett. **20**, 807-809 (1995)

2.48   F.A. Jenkins, H. E. White: *Fundamental of Optics*, 4th edn. (McGraw-Hill, New York 1976) pp.166-167

2.49   W.H. Knox, J.E. Henry, K.W. Goossen, K.D. Li, B. Tell, D.A.B. Miller, D.S. Chemla, A.C. Gossard, J. English, S. Schmitt-Rink, IEEE J. QE-**25**, 2586-2595 (1989)

2.50   W. Sha, T.B. Norris, J.W. Burm, D. Woodard, W.J. Schaff: New coherent detector for terahertz radiation based on electroabsorption, Appl. Phys. Lett. **61**, 1763-1765 (1992)

2.51   G.W. Chantry, H.M. Evans, J.W. Fleming, H.A. Gebbie: TPX, a new materal for optical components in the far infra-red spectral region, Infrared Phys. **9**, 31-33 (1969)

2.52   J.M. Chwalek, J.F. Whitaker, G.A. Mourou: Submillimetre wave response of superconducting $YBa_2 Cu_3 O_{7-x}$ using coherent time-domain spectroscopy, Electron. Lett. **27**, 447-8 (1991)

2.53   M. van Exter, D. Grischkowsky: Optical and electronic properties of doped silicon from 0.1 to 2 THz, Appl. Phys. Lett. **56**, 1694-1696 (1990)

2.54   M.C. Nuss, K.W. Goossen, J.P. Gordon, P.M. Mankiewich, M.L. O'Malley, M. Bhushan: Terahertz time-domain measurement of the conductivity and superconducting band gap in niobium, J. Appl. Phys. **70**, 2238-41 (1991)

2.55   M.V. Klein: *Optics* (Wiley, New York 1970) pp.582-585

2.56   H. Roskos, M.C. Nuss, J. Shah, B. Tell, J. Cunningham: Terahertz absorption between split subbands in coupled quantum wells, in *Picosecond Electronics and Optoelectronics IX* (Salt Lake City, UT, USA 1991) (OSA, Washington, DC 1991) pp.24-30

2.57   W.P. Chen, Y.J. Chen, E. Burstein: The interface EM modes of a "surface quantized" plasma layer on a semiconductor surface, Surf. Sci. **58**, 263-265 (1976)

2.58   K.L. Kliewer, R. Fuchs: Collective electronic motion in a metallic slab, Phys. Rev. **153**, 498-512 (1967)

2.59   S.E. Ralph, S. Perkowitz, N. Katzenellenbogen, D. Grischkowsky: Terahertz spectroscopy of optically thick multilayered semiconductor structures, J. Opt. Soc. Am. B **11**, 2528-2532 (1994)

2.60   D. Grischkowsky, S. Keiding, M. van Exter, C. Fattinger: Far-infrared time-domain spectroscopy with terahertz beams of dielectrics and semiconductors, J. Opt. Soc. Am. B **7**, 2006-2015 (1990)

2.61   D. Grischkowsky, S. Keiding: THz time-domain spectroscopy of high-$T_c$ substrates, Appl. Phys. Lett. **57**, 1055-1057 (1990)

2.62   M. Ree, K.J. Chen, D.P. Kirby, N. Katzenellenbogen, D. Grischkowsky: Anisotropic properties of high-temperature polyimide thin films: dielectric and thermal-expansion behaviors, J. Appl. Phys. **72**, 2014-2021 (1992)

2.63   S.E. Ralph, D. Grischkowsky: THz spectroscopy and source characterization by optoelectronic interferometry, Appl. Phys. Lett. **60**, 1070-1072 (1992)

2.64   M. van Exter, D. Grischkowsky: Carrier dynamics of electrons and holes in moderately doped silicon, Phys. Rev. B **41**, 12140-12149 (1990)

2.65   N. Katzenellenbogen, D. Grischkowsky: Electrical characterization to 4 THz of N- and P-type GaAs using THz time-domain spectroscopy, Appl. Phys. Lett. **61**, 840-2 (1992)

2.66   W.J. Walecki, D. Some, V.G. Kozlov, A.V. Nurmikko: Terahertz electromagnetic transients as probes of a two-dimensional electron gas, Appl. Phys. Lett. **63**, 1809-1811 (1993)

2.67   D. Some, A. V. Nurmikko: Coherent transient cyclotron emission from photoexcited GaAs, Phys. Rev. B **50**, 5783-5786 (1994)

2.68   D. Some, A.V. Nurmikko: Real-time electron cyclotron oscillations observed by terahertz techniques in semiconductor heterostructures, Appl. Phys. Lett. **66**, 1-3 (1995)

2.69   D.A.B. Miller, D.S. Chemla, S. Schmitt-Rink: Phys. Rev. B **33**, 6676-6982 (1986)

2.70   T. Ando, A.B. Fowler, F. Stern: Rev. Mod. Phys. **54**, 437-672 (1982)

2.71   E. Yablonovitch: Phys. Rev. Lett. **58**, 2059-2061 (1987)

2.72   W.M. Robertson, G. Arjavalingam, R. D. Meade, K. D. Brommer, A. M. Rappe, J. D. Joannopoulos: Measurement of photonic band structure in a two-dimensional periodic dielectric array, Phys. Rev. Lett. **68**, 2023-6 (1992)

2.73   W.M. Robertson, G. Arjavalingam, R.D. Meade, K.D. Brommer, A.M. Rappe, J.D. Joannopoulos: Measurement of the photon dispersion relation in two-dimensional ordered dielectric arrays, J. Opt. Soc. Am. B **10**, 322-327 (1993)

2.74   S.Y. Lin, G. Arjavalingam: Tunneling of electromagnetic waves in two-dimensional photonic crystals, Opt. Lett. **18**, 1666-1668 (1993)

2.75   S.Y. Lin, G. Arjavalingam: Photonic bound states in two-dimensional photonic crystals probed by coherent-microwave transient spectroscopy, J. Opt. Soc. Am. B **10**, 2124-2127 (1994)

2.76   E. Ozbay, E. Michel, G. Tuttle, R. Biswas, K.M. Ho, J. Bostak, D.M. Bloom: Terahertz spectroscopy of three-dimensional photonic band-gap crystals, Opt. Lett. **19**, 1155-1157 (1994)

2.77   E. Ozbay, A. Abeyta, G. Tuttle, M. Tringides, R. Biswas, C.T. Chan, C.M. Soukoulis, K.M. Ho: Measurement of a three-dimensional photonic band gap in a crystal structure made of dielectric rods, Phys. Rev. B **50**, 1945-1948 (1994)

2.78   R.E. Glover, M. Tinkhma: Conductivity of superconducting films for photon energies between 0.3 and 40 $kT_c$, Phys. Rev. **108**, 243-256 (1957)

2.79    L.H. Palmer, M. Tinkham: Far-infrared absorption of thin superconducting lead films, Phys. Rev. **165**, 588-595 (1968)

2.80    M.C. Nuss, P.M. Mankiewich, M.L. O'Malley, E.H. Westerwick, P.B. Little-wood: Dynamic conductivity and 'coherence peak' in $YBa_2 Cu_3 O_7$ superconductors, Phys. Rev. Lett. **66**, 3305-8 (1991)

2.81    C.M. Varma, P.B. Littlewood, S. Schmitt-Rink, E. Abrahams, A.E. Ruckenstein, Phys. Rev. Lett. **63**, 1996-1999 (1989)

2.82    P.B. Littlewood, C.M. Varma: J. Appl. Phys. **69**, 4979-4984 (1991)

2.83    F. Gao, J.F. Whitaker, Y. Liu, C. Uher, C.E. Platt, M.V. Klein: Terahertz transmission of a $Ba_{1-x} K_x BiO_3$ film probed by coherent time-domain spectroscopy, Phys. Rev. B **52**, 3607-13 (1995)

2.84    R. Buhleier, S.D. Brorson, I.E. Trofimov, J.O. White, H.U. Habermeier, J. Kuhl: Anomalous behavior of the complex conductivity of $Y_{1-x} Pr_x Ba_2 Cu_3 O_7$ observed with THz spectroscopy, Phys. Rev. B **50**, 9672-9675 (1994)

2.85    S. Spielman, B. Parks, J. Orenstein, D.T. Nemeth, F. Ludwig, J. Clarke, P. Merchant, D.J. Lew: Observation of the quasiparticle Hall effect in superconducting $YBa_2 Cu_3 O_{7-\delta}$, Phys. Rev. Lett. **73**, 1537-1540 (1994)

2.86    B. Parks, S. Spielman, J. Orenstein, D.T. Nemeth, F. Ludwig, J. Clarke, P. Merchant, D.J. Lew: Phase-sensitive measurements of vortex dynamics in the terahertz domain, Phys. Rev. Lett. **74**, 3265-3268 (1995)

2.87    B.I. Greene, P.N. Saeta, D.R. Dykaar, S. Schmitt-Rink, S.L. Chuang: Far-infrared light generation at semiconductor surfaces and its spectroscopic applications, IEEE J. QE-**28**, 2302-2312 (1992)

2.88    M.C. Nuss: Time-resolved terahertz conductivity of photoinjected hot electrons in gallium arsenide, in *Ultrafast Phenomena VI*, ed. by T. Yajima, K. Yoshihara, C.B. Harris, S. Shionoya, Springer Ser. Chem. Phys., Vol.48 (Springer, Berlin, Heidelberg 1988) pp.215-217

2.89    P.N. Saeta, J.F. Federici, B.I. Greene, D.R. Dykaar: Intervalley scattering in GaAs and InP probed by pulsed far-infrared transmission spectroscopy, Appl. Phys. Lett. **60**, 1477-1479 (1992)

2.90    R.H.M. Groeneveld, D. Grischkowsky: Picosecond time-resolved far-infrared experiments on carriers and excitons in GaAs-AlGaAs multiple quantum wells, J. Opt. Soc. Am. B **11**, 2502-2507 (1994)

2.91    C. Karadi, S. Jauhar, L. P. Kouwenhoven, K. Wald, J. Orenstein, P.L. McEuen, Y. Nagamune, H. Sakaki: Dynamic response of a quantum point contact, J. Opt. Soc. Am. B **11**, 2566-2571 (1994)

2.92    S. Verghese, C. Karadi, C.A. Mears, J. Orenstein, P.L. Richards, A.T. Barfknecht: Broadband response of the quasiparticle current in a superconducting tunnel junction, Appl. Phys. Lett. **64**, 915-917 (1994)

2.93    L.P. Kouwenhoven, S. Jauhar, J. Orenstein, P.L. McEuen, Y. Nagamune, J. Motohisa, H. Sakaki: Observation of photon-assisted tunneling through a quantum dot, Phys. Rev. Lett. **73**, 3443-3446 (1994)

2.94    J.R. Tucker, M.J. Feldman: Rev. Mod. Phys. **57**, 1055-1113 (1985)

2.95    Y.R. Shen: *The Principles of Nonlinear Optics* (Wiley, New York 1984)

2.96   C. Changsungsan, L. Tsang, S.L. Chuang: Coherent terahertz emission from coupled quantum wells with exciton effects, J. Opt. Soc. Am. B 11, 2508-2518 (1994)

2.97   B.B. Hu, X.C. Zhang, D.H. Auston: Temperature dependence of femtosecond electromagnetic radiation from semiconductor surfaces, Appl. Phys. Lett. 57, 2629-2631 (1990)

2.98   J.H. Son, T.B. Norris, J.F. Whitaker: Terahertz electromagnetic pulses as probes for transient velocity overshoot in GaAs and Si, J. Opt. Soc. Am. B 11, 2519-2527 (1994)

2.99   B.B. Hu, X.C. Zhang, D.H. Auston, P.R. Smith: Free-space radiation from electro-optic crystals, Appl. Phys. Lett. 56, 506-508 (1990)

2.100  L. Xu, X.C. Zhang, D.H. Auston: Terahertz beam generation by femtosecond optical pulses in electro-optic materials, Appl. Phys. Lett. 61, 1784-1786 (1992)

2.101  X.C. Zhang, Y. Jin, X.F. Ma: Coherent measurement of THz optical rectification from electro-optic crystals, Appl. Phys. Lett. 61, 2764-2766 (1992)

2.102  T. Dekorsky, H. Auer, C. Waschke, H.J. Bakker, H.G. Roskos, H. Kurz, V. Wagner, P. Grosse: Emission of submillimeter electromagnetic waves by coherent phonons, Phys. Rev. Lett. 74, 738-741 (1995)

2.103  X.C. Zhang, X.F. Ma, Y. Jin, T.M. Lu, E.P. Boden, P.D. Phelps, K.R. Stewart, C.P. Yakymyshyn: Terahertz optical rectification from a nonlinear organic crystal, Appl. Phys. Lett. 61, 3080-3082 (1992)

2.104  H.G. Roskos, M.C. Nuss, J. Shah, K. Leo, D.A. B. Miller, A.M. Fox, S. Schmitt-Rink, K. Kohler: Coherent submillimeter-wave emission from charge oscillations in a double-well potential, Phys. Rev. Lett. 68, 2216-2219 (1992)

2.105  P.C. M. Planken, M.C. Nuss, I. Brener, K.W. Goossen, M.S.C. Luo, S.L. Chuang, L. Pfeiffer: Terahertz emission in single quantum wells after coherent optical excitation of light hole and heavy hole excitons, Phys. Rev. Lett. 69, 3800-3803 (1992)

2.106  P.C.M. Planken, I. Brener, M.C. Nuss, M.S.C. Luo, S.L. Chuang: Coherent control of terahertz charge oscillations in a coupled quantum well using phase-locked optical pulses, Phys. Rev. B 48, 4903-4906 (1993)

2.107  P.C.M. Planken, I. Brener, M.C. Nuss, M.S.C. Luo, S.L. Chuang, L.N. Pfeiffer: THz radiation from coherent population changes in quantum wells, Phys. Rev. B 49, 4668-4672 (1994)

2.108  M.C. Nuss, P.C.M. Planken, I. Brener, H.G. Roskos, M.S.C. Luo, S.L. Chuang: Terahertz electromagnetic radiation from quantum wells, Appl. Phys. B 58, 249-259 (1994)

2.109  I. Brener, P.C.M. Planken, M.C. Nuss, M.S.C. Luo, C.S. Lien, L. Pfeiffer, D.E. Leaird, A.M. Weiner: Coherent control of terahertz emission and carrier populations in semiconductor heterostructures, J. Opt. Soc. Am. B 11, 2457-2469 (1994)

2.110  C. Waschke, H.G. Roskos, R. Schwedler, K. Leo, H. Kurz, K. Kohler: Coherent submillimeter-wave emission from Bloch oscillations in a semiconductor superlattice, Phys. Rev. Lett. 70, 3319-3322 (1993)

2.111  H.G. Roskos, C. Waschke, R. Schwedler, P. Leisching, Y. Dhaibi, H. Kurz, K. Köhler: Bloch oscillations in GaAs/AlGaAs superlattices after excitation well above the bandgap, Superlattices and Microstructures 15, 281-285 (1994)

2.112  B.B. Hu and M.C. Nuss: Imaging with terahertz waves, Opt. Lett. **20**, 1716-1718 (1995)

2.113  M.C. Nuss: Chemistry is Right for T-Ray Imaging, in IEEE Circuits and Devices **12**, 25-30 (1996)

2.114  M.J.W. Rodwell, S.T. Allen, R.Y. Yu, M.G. Case, U. Bhattacharya, M. Reddy, E. Carman, M. Kamegawa, Y. Konishi, J. Pusl, R. Pullela: Active and nonlinear wave propagation devices in ultrafast electronics and optoelectronics, Proc. IEEE **82**, 1037-1059 (1994)

2.115  Y. Konishi, M. Kamegawa, M. Case, R. Yu, M.J.W. Rodwell, R.A. York, D.B. Rutledge: Picosecond electrical spectroscopy using monolithic GaAs circuits, Appl. Phys. Lett. **61**, 2829-2831 (1992)

2.116  J.S. Bostak, D.W. van der Weide, D.M. Bloom, B.A. Auld: All-electronic terahertz spectroscopy system with terahertz free-space pulses, J. Opt. Soc. Am. B **11**, 2561-2565 (1994)

# 3. Coherent Source Submillimeter Wave Spectroscopy

Gennadi Kozlov and Alexander Volkov
With 40 Figures

## List of Symbols

| | |
|---|---|
| $\lambda$ | Wavelength of radiation |
| $f$ | Technical (source) frequency |
| $\omega$ | Circular frequency |
| $\nu$ | Spectroscopic frequency |
| $\Delta f, \Delta \nu$ | Linewidth |
| $q$ | Wave vector |
| $E$ | Electric field |
| $U$ | Voltage |
| $I$ | Intensity of radiation |
| $T$ | Transmissivity |
| $R$ | Reflectivity |
| $A$ | Absorptivity |
| $\varphi$ | Phase shift in transmission |
| $\Psi$ | Phase shift in reflection |
| $N = n + \mathrm{i}k$ | Refractive index |
| $\varepsilon = \varepsilon' + \mathrm{i}\varepsilon''$ | Dielectric constant |
| $\sigma = \sigma' + \mathrm{i}\sigma''$ | Conductivity |
| $\sigma_0$ | $dc$ Conductivity |
| $\varepsilon_0$ | Static permittivity |
| $\varepsilon_\infty$ | Optical permittivity |
| $\Delta \varepsilon$ | Dielectric contribution |
| $S$ | Oscillator (relaxator) strength |
| $\tau$ | Relaxation time |
| $\nu_0$ | Resonance frequency |
| $\gamma$ | Damping |

## Reductions

| | |
|---|---|
| BWO | Backward Wave Oscillator |
| SBMM | SuBMilliMeter |
| MW | MicroWave |
| MM | MilliMeter |
| IR | Infrared |

Topics in Applied Physics, Vol. 74
**Millimeter and Submillimeter Wave Spectroscopy of Solids** Ed.: G. Grüner
© Springer-Verlag Berlin Heidelberg 1998

In this review we describe our experience in the application of tunable monochromatic generators – *Backward Wave Oscillators* (BWOs) sometimes called *carcinotrons* (from the Greek "crawfish", creeping back) – to submillimeter radiophysical and dielectric measurements ($\lambda \approx$ 3–0.3 mm). Along with the klystrons, magnetrons and traveling wave oscillators, BWOs are classical electrovacuum microwave generators (sort of traveling wave oscillators), possessing two important distinguishing abilities: to generate extremely short wavelengths – down to $\lambda \approx 0.2$ mm, and to electronically tune the working frequency in a broad range – up to $\pm$ 30% from the central value.

Millimeter and submillimeter BWOs arose as a result of painstaking work of scientists and engineers, who after World War II realized the importance of expanding microwave electronics towards the short-wavelength part of the spectrum, from the centimeter to millimeter wavelength range. Major progress occurred in the 1960s. In the USSR and France, in particular, sets of short-wavelength BWOs were developed and brought to industrial production, available for use in technology and science [3.1,2] Today one can read about these results in reviews [3.3–5].

In the succeeding 30 years no major improvements were made in the development of short-wave BWOs. In millimeter technology, from the long-wavelength side, semiconductor generators based on avalanche (IMPATT) and Gunn diodes began to actively force out BWOs, because they are more compact, economical, and convenient to operate. Industrial production of short-wave BWOs with $\lambda < 1.5$ mm was reduced, too, and seems to have survived only in Russia.

Meanwhile, BWOs are still beyond competition in solving a large number of physical and technological problems. We demonstrate this below by SBMM dielectric measurements done with BWOs. Some results of these measurements cannot be obtained at present by any other technique. On the basis of BWOs we have developed measurement techniques for the $\lambda \approx$ 3–0.3 mm domain and have studied a) electrodynamic properties of various quasi-optical devices, and b) dielectric properties of a wide range of substances. The methods developed are of a hybrid type, combining elements of both microwave technology and infrared spectroscopy: on the one hand they use the high quality radiation – high intensity ($\approx 10$ mW), high monochromaticity ($\Delta f / f \approx 10^{-5}$), high degree of polarization (99.99 %) – and on the other hand use open space quasi-optical measuring schemes, and continuous tuning of the operating frequency over wide ranges. As a result, information of the highest quality is registered rapidly in a real-time scale, yielding spectra of absolute values of both parts (real and imaginary) of the dielectric function (permittivity or conductivity: $\varepsilon'$ and $\varepsilon''$, $\sigma'$ and $\sigma''$).

High productivity of BWO-spectrometers allowed us to perform measurements on thousands of samples – single crystals and ceramics, glasses and polymers, powders, composites, liquids, films, fibers, etc., – and to compile reference material on the dielectric properties of microwave and optical materials. We have investigated fundamental regularities of the frequency-temperature behavior of the dielectric response function in substances of different classes – simple dielectrics, ferroelectrics, ionic conductors, dipole glasses, incommensurate crystals, semiconductors, superconductors, low-dimensional conductors, and antiferromagnets.

## 3.1 Submillimeter Quasi-optical Technique

### 3.1.1 Backward Wave Oscillators

BWOs are miniature electrovacuum devices fitted in a metal casing (Fig. 3.1). Put in a magnetic field and supplied with a high voltage, the BWO emits monochromatic electromagnetic radiation with a characteristic output power of $\approx 10\,$mW. The radiation emitted into free space can easily be detected by a far-infrared detector. Radiation of long-wavelength BWOs, whose output power reaches 200 mW, can be felt by the hand at the edge of the waveguide.

Figure 3.2 shows schematically the arrangement of the BWO, which essentially amounts to an electrovacuum diode or triode. While the heater (1) is switched on, the cathode (2) (electron gun) emits electrons (3) which, accelerated by a high voltage electrical field, travel in a vacuum toward the anode (4) (collector). Collimated in a beam by an external magnetic field [magnet (5)] the electrons fly over a comb-like fine-structure electrode (6) (slowing system) intended to transfer the kinetic energy of the electrons to the electromagnetic field. Actually, moving in the variable potential of the slowing system, the electrons are grouped periodically in bunches (velocity/phase modulation) and form an electromagnetic wave (7) traveling in the opposite direction to the electrons (backward wave). This radiation comes out through an oversize waveguide (8). The velocity of the electrons, and thus the radiation frequency, are determined by the magnitude of the accelerating field.

The BWO has to be adjusted in the magnet by rotating it around two axes, one directed along the waveguide and the other orthogonal to it and the

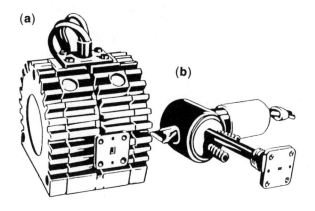

**(a)**

**(b)**

**Fig. 3.1 a,b.** Backward wave oscillators of Russian production: **(a)** – packetized (inside magnet), **(b)** – unpacketized (bare)

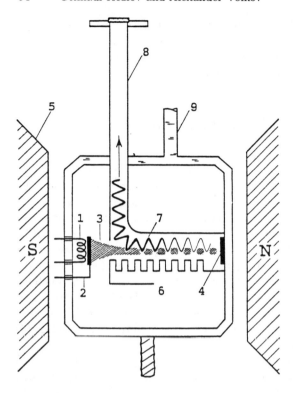

**Fig. 3.2.** Schematic diagram of backward wave oscillator: *1* – heater, *2* – cathode, *3* – electron beam, *4* – collector (anode), *5* – permanent magnet, *6* – slowing system, *7* – electromagnetic wave, *8* – waveguide, *9* – water cooling

magnetic field. Typical requirements on the adjustment are about 1 degree; they rapidly become more stringent for the short wavelength BWOs. In spite of looking simple in construction, BWOs are highly sophisticated devices, working in an extremely intensive mode. The reason is that scaling down the BWO to short wavelengths imposes very arduous conflicting requirements on the electrical and geometrical parameters of these devices. Designers had to combine a large number of small gaps (millimeters and fractions of millimeters) with high voltages (up to 6.5 kV), high temperatures (up to 1200°C on the cathode) and high vacuum (up to $10^{-8}$ Torr). Unprecedented accuracies were needed to fabricate slowing systems (up to 200 rods separated by $\approx 10\,\mu$m) and position the electrodes. The emitting efficiency of cathodes and the electron current density in the beam had to be increased many times – up to 15 A/cm$^2$ and 150 A/cm$^2$, respectively.

BWOs can be divided into two main types according to their construction – those having their own magnet (packetized) and those without it (unpacketized) (Fig. 3.1). Low-frequency BWOs (with $f < 180$ GHz) are mostly produced in a packetized form. In this variant the body of BWO is irreversibly bricked up in a small samarium-cobalt magnet. The whole construction is very compact, weighing 1 kg, with a developed finned surface, and is built to be cooled by air.

BWOs with working frequencies $f > 180\,\text{GHz}$ are produced unpacketized since they require stronger magnetic fields. When going from 180 to 1200 GHz the magnetic field strength increases from 6 to 12 kOe (in a gap of $\approx 30\text{--}35\,\text{mm}$). These conditions can be realized in massive Sm–Co or NdFeB magnetic systems (weighing 10 kg or more) or in electromagnets (100 kg). To mount the BWOs in such magnets requires a high precision mechanical adjustment system, designed individually for a certain type of magnet. The body of unpacketized BWOs is cooled by water (9).

Figure 3.3 shows the radiation capabilities of the BWOs. Both the frequency and output power of the BWOs change depending on the high voltage applied. Typically the output frequency $f$ of all BWOs depends on the high voltage according to the relation $f \propto U^{1/2}$. At the same time the output power $vs$ voltage $I(U)$ (BWO's spectral pattern) looks like a random function, unique for each BWO but is highly reproducible. The BWO output power may change by a factor of tens between the minima and maxima of

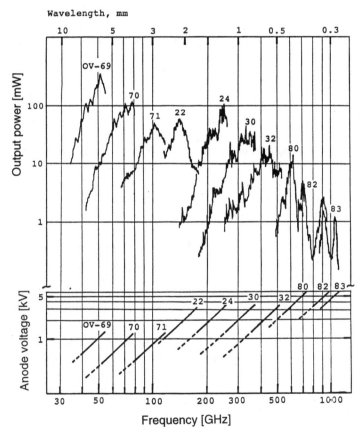

**Fig. 3.3.** Panorama of radiation characteristics of BWOs: output power and operating frequencies $vs$ anode voltage measured in free space

the BOW's spectral pattern. For all BWOs the most intensive part is distinguished in the spectral pattern $I(U)$, usually indicated in the specifications as the working range. A low-frequency tail is neglected in specifications as a nonoperational part. However, our experience has shown that this part is very useful in spectroscopy, providing reliable frequency linking and overlapping of experimental data obtained with different BWOs. The nonoperational part reveals a frequency tuning range comparable with the tuning range of the working part.

From the spectroscopic viewpoint, the most important question concerns the properties of the instrumental generation line of the BWO – its width and intensity, stability and reproducibility, speed, step and range of frequency tuning. In general, the smaller the linewidth, the higher the stability and sweep rapidity, the wider the tuning range, the more complicated is the problem of instrumental line forming. Strictly speaking, the problem relates not only to the BWO, but to the entire magnet-BWO-power supply system, since the properties of the instrumental line directly depend on the BWO operation conditions – on properties of feeding voltages (anode and partly heater), and stability of the magnetic field. While the parameters of industrially produced BWOs are more or less standardized, there are, at present, no commercially available universal power supplies and magnets for BWOs. This means that in all BWO applications the problem of the development of the instrumental line is solved in a particular spot in accordance to the specific requirements of the job. As shown below in the description of the dielectric measurements on the BWO spectrometer, without phase-lock control, the monochromaticity of the BWO radiation and its frequency reproducibility under high voltage tuning conditions can be maintained at the level of $\Delta f \approx 10^{-5} f$. The short-term stability of the radiation power and its reproducibility are up to $\Delta I \approx 3 \cdot 10^{-3} I$.

Power supplies providing the above parameters in our BWO-spectrometers are universal for all BWOs. They are constructed according to a classical scheme based on a control tube and give stabilized computer-tuned voltages up to 6.5 kV, with operating currents up to 60 mA. Noise, ripples, and short-term deviation of the voltage in the working regime do not exceed 20 mV. The high rapidity of the power supplies is nontrivial: the setting time of the high voltage is about $10^{-2}$ s. Principally, the inertia of the BWOs frequency tuning does not exceed $10^{-8}$ s [3.3]. In addition to the anode voltage, feeding of the heater of the BWO is also stabilized. Due to construction features of the BWOs (grounded anode), the heater supplying circuit have to be under high voltage. This makes a universal power supply a rather complicated and expensive device. The problem of supplying power to the BWO is considerably simplified in the case of packetized BWOs. The small size of the magnet allows a packetized BWO to be placed inside the electronic block and left under high voltage.

## 3.1.2 Submillimeter BWO-Spectrometer "Epsilon"

Among the variety of BWO-based measuring installations known from literature [3.6–9] the spectrometer "Epsilon" described below is distinguished by two features: a rapid, reproducible, high precision, wide-range scanning of the frequency and unique equipment for dielectric measurements [3.10–12]. In our spectrometers the BWO acts as a source-monochromator block in the structure of a classical grid infrared spectrometer, producing a narrow tunable instrumental line of radiation and illuminating the sample under investigation. The radiation detector registers the amplitude spectra (frequency dependencies) of the transmitted or reflected signal.

**a) Apparatus.** "Epsilon" consists of five main subsystems: the generator, the registration system, the optical measuring channel, the unit for holding, adjusting, and thermal variation of the samples, and the control system.

The first subsystem contains replaceable BWOs, the electromagnet, and the power supply for the BWOs.

The registration system consists of a detecting cell (room temperature optical-acoustical Golay cell or cooled bolometer), a 25 Hz radiation modulator, an amplifier with a synchronous detector, and a sampling-storage-reset circuit. Like the BWO power supply control, the operation of this unit is also synchronized with a modulation frequency. The dynamic range of the electronic registration unit is $10^4$. Added to the dynamical range of the quasi-optical attenuator ($10^2$–$10^3$) it provides a value of $10^6$–$10^7$ for the whole BWO-spectrometer.

The measuring section of the spectrometer is, in essence, an optical measuring channel, it is highly flexible and can be changed by simple replacement of elements installed on optical rails and carriers (Fig. 3.4). While the specific configuration of this unit depends on the choice of the experimental method, the basic invariable principles of its construction are the following: the radiation propagates in free space; a beam with a diameter $\approx 40$ mm is formed by dielectric lenses or by metallic parabolic mirrors; plain one-dimensional wire grids with a period $L \ll \lambda$ are used as polarizers and beam splitters; the sample is placed in a channel tightly pressed against a metallic diaphragm.

The sample temperature can be varied from 1.6 to 1000 K. For this purpose, we use optical cryostats and thermostats which are specially developed for BWO spectrometers. They are equipped with large inclined windows made of thin polymer films in order to avoid standing waves in the channel.

The operation of the spectrometer is controlled by a computer.

**b) Idea of Dielectric Measurements.** In all the techniques worked out by us, the radiation is normally incident on the sample which is prepared in the form of a plane-parallel plate. Depending on the method used, the measurable quantities can either be the transmission coefficient $T$ of the sample and the phase shift $\phi$ of the transmitted wave, or the parameters $R$ and $\psi$, the reflection coefficient and the reflected wave phase shift, corresponding to the

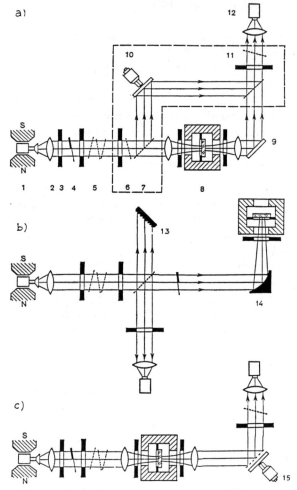

**Fig. 3.4 a–c.** Quasi-optical measuring channel of the BWO-spectrometer "Epsilon": (**a**) – dielectric measurements via $T(f)$ and $\phi(f)$ spectra, (**b**) – reflectivity measurement, (**c**) – birefringence measurement, $1$ – BWO, $2$ – dielectric lenses, $3$ – absorbing diaphragms, $4$ – chopper, $5$ – grid (metallic film) attenuator, $6$ – polarizer, $7$ – beam splitter, $8$ – thermostat with the sample, $9$ – mirror (phase modulator), $10$ – phase-shift compensator, $11$ – analyzer, $12$ – detector, $13$ – absorber, $14$ – parabolic mirror, $15$ – phase transducer

bulk sample. The mathematical relationships connecting these measurable quantities with the dielectric parameters of material $\varepsilon'$ and $\varepsilon''$ are well-known in optics [3.13]:

$$T = e^{-\frac{4\pi k d}{\lambda}} \frac{(1-R)^2 + 4R\sin^2\psi}{\left(1 - Re^{-\frac{4\pi k d}{\lambda}}\right)^2 + 4Re^{-\frac{4\pi k d}{\lambda}}\sin^2\left(\frac{2\pi n d}{\lambda} + \psi\right)} \tag{3.1}$$

$$\phi = \frac{2\pi n d}{\lambda} - \arctan \frac{k(n^2 + k^2 - 1)}{(k^2 + n^2)(2 + n)n}$$

$$+ \arctan \frac{Re^{-\frac{4\pi k d}{\lambda}} \sin 2\left(\frac{2\pi n d}{\lambda} + \psi\right)}{1 - Re^{-\frac{4\pi k d}{\lambda}} \cos 2\left(\frac{2\pi n d}{\lambda} + \psi\right)}, \tag{3.2}$$

$$R = \frac{(n-1)^2 + k^2}{(n+1)^2 + k^2}, \quad \psi = \arctan\left(\frac{2k}{n^2 + k^2 - 1}\right), \tag{3.3}$$

$$\varepsilon' = n^2 - k^2, \quad \varepsilon'' = 2nk, \tag{3.4}$$

where $d$ is the sample thickness, and $n$ and $k$ are the refractive and extinction coefficients, respectively, i.e., the optical parameters of the material.

Obviously, any pair of quantities from $T$, $\phi$, $R$ and $\psi$ can be used to calculate $n$ and $k$ ($\varepsilon'$ and $\varepsilon''$). Besides, transmission data obtained for samples with different thicknesses $T_1$ and $T_2$ is sometimes used for dielectric measurements. In far-infrared spectroscopy, where transmission measurements are difficult, the quantities $R$ and $\psi$ are used as a rule. In general, the choice of experimental methods is quite limited and uniquely determined by the actual technical possibilities of measuring $T$, $\phi$, $R$ and $\psi$. Meanwhile, it is clear that different pairs of measurable quantities are far from being equivalent in providing the maximal accuracies in the values $n$ and $k$. In view of the considerable nonlinearity of (3.1–4), the uncertainties in the values of $n$ and $k$ depend on both the errors in the measurement of $T$, $\phi$, $R$, $\psi$, and the absolute values of $n$ and $k$ themselves.

While developing the BWO methods of dielectric measurements we have analyzed various ways of determining $n$ and $k$, depending on the errors in BWO measurement of $T$, $\phi$, $R$, and $\psi$ [3.14]. It turned out that under nearly identical conditions, the $T$ and $\phi$ method gives the best results. The $T$ and $\phi$ method is the most preferable in our dielectric measurements and is always used when the radiation power is sufficient to penetrate through the sample and be registered by the detector.

c) **Dielectric Measurement Procedure.** Figure 3.4 shows a set of measuring schemes of the "Epsilon" spectrometer: for $T$, $\phi$, $R$, and birefringence measurements. The first $(T, \phi)$-configuration is a basic one. The $(T\ vs\ f)$ and $(\phi\ vs\ f)$ spectra are recorded separately in two stages. First the $(T\ vs\ f)$ spectrum is recorded with the help of a simple "transmission" geometry, and the section isolated in Fig. 3.4 by the dashed line is not used. The generator operates in the frequency scanning mode and the scanning is carried out at different points by varying the BWO supply voltage in steps. The signal at the detector is recorded as a function of the radiation frequency. In order to eliminate the instrumental function of the spectrometer, the procedure is repeated twice, once with the sample in the channel and once without it (reference). The transmission spectrum $T(f)$ is calculated by a termwise division of the two data files.

The second stage involves the recording of the $(\phi\ vs\ f)$ spectrum. All the elements of the quasi-optical channel are now used, and form a Rozhdestven-skii two-beam polarization interferometer (Mach–Zehnder interferometer). The interferometer operates in such a way that during the frequency scan, the mobile mirror (10) moves all the time, controlled by a feedback tracing system in order to sustain the interferometer in a balanced state (zero signal on the detector). The measurable quantity in this case is the displacement $\Delta(f)$ of this mobile mirror. In other words, the spectrometer registers the change in the optical thickness of the sample $vs$ frequency. As in the case of recording the $(T\ vs\ f)$ spectrum, the basic measurement in this case (with the sample in the channel) is also preceded by a reference, i.e., by the measurement of $\Delta_0(f)$ of an empty channel. The phase spectrum $\phi(f)$ of the sample is determined from the difference $\Delta(f) - \Delta_0(f)$.

Clearly, because of the complex instrumental function of the BWO spectrometer, the accuracies of the $(T\ vs\ f)$ and $(\phi\ vs\ f)$ measurements strongly depend on the frequency matching of the sample and reference spectra. The reproducibility of the results in our spectrometers is 0.1–1 % (depending on the scanning speed) for $T(f)$ measurements and $\approx 1\,\mu m$ for $\Delta(f)$ measurements. These values represent a typical scatter of the data for $T$ and $\Delta$ at individual frequency points upon a comparison of several successively recorded spectra.

Figure 3.5 illustrates the described stages: $T(f)$ and $\phi(f)$ measurements and corresponding calculated $\varepsilon'(f)$ and $\varepsilon''(f)$ spectra. This example refers to a very thin (0.1 mm) plate of a semiconducting TlGaSe$_2$ crystal. In view of the considerable absorption in TlGaSe$_2$, even a very thin sample significantly weakens the radiation: $T \approx 10^{-4}$, note the logarithmic scale. Here we used the Golay cell as a detector and the $T(f)$ spectrum shows that the Golay cell coupled with BWOs can be confidently used for transmissivities as low as $T \approx 10^{-5}$. This limit can be further lowered by about two orders of magnitude by using cooled detectors. However, it has to be noted that while working with the small signals a complicated problem arises of suppressing the parasitic coherent signal at the detector [3.12].

Assuming that the bulk sample thicknesses can be reduced to a few tens of microns, then values of $\varepsilon'$, $\varepsilon'' \approx 1000$ ($\sigma \approx 100\,\Omega^{-1}\ cm^{-1}$) are accessible for measurements on our BWO spectrometers in the simple transmission geometry. The average accuracy of such dielectric measurements over a wide frequency range is $\approx 5\,\%$ for $\varepsilon'$ and $\approx 10\,\%$ for $\varepsilon''$.

The described method is universal. It guarantees obtaining the $\varepsilon(f)$ or $\sigma(f)$ spectra as long as the sample is penetrable for radiation at the level of $T > 10^{-4}$. This condition is fulfilled in the vast majority of the practically important cases.

For transparent samples the same data can be obtained with a more simple and rapid method. Figure 3.6 presents transmission spectra of a number

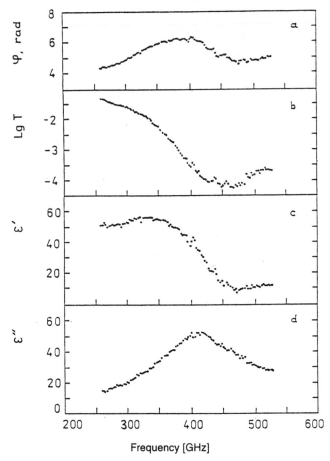

**Fig. 3.5 a–d.** Dielectric measurements in the tuning range of one BWO (O-32): (a) and (b) – $T(f)$ and $\phi(f)$ spectra of dielectric plate (TlGaSe$_2$ of 0.1 mm thick), (c) and (d) – $\varepsilon'(f)$ and $\varepsilon''(f)$ spectra calculated from $T(f)$ and $\phi(f)$ spectra. The whole cycle is 5 min long and contains 100 experimental points

of dielectric plates prepared from different transparent materials. The characteristic feature of the spectra is a periodic transmission pattern caused by the interference of the radiation inside the plates (compare with the smooth $T(f)$ spectrum in Fig. 3.5b). As follows from (3.1) the period of the oscillations relates directly to the refraction index $n$:

$$\frac{m\lambda}{2} = nd,$$

where $\lambda$ is the radiation wavelength, $m$ is the interference maximum number, and $d$ is the thickness of the plate. The dielectric loss in the substance determines the amplitudes of the maxima and the swing of the oscillations. In the case under discussion, a plane-parallel sample operates as a multibeam

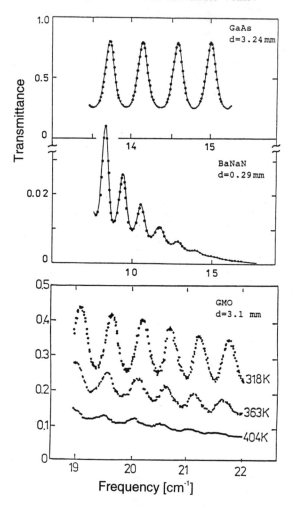

**Fig. 3.6.** Typical SBMM transmission spectra of plane-parallel dielectric plates. *Points* – experiment, *lines* – fit with (3.1)

dielectric-filled interferometer that allows for the determination of its dielectric parameters. We evaluate $\varepsilon'(f)$ and $\varepsilon''(f)$ spectra by fitting the theoretical oscillating curve $T(f)$ given by (3.1) to the experimental spectrum $T(f)$. Here the frequency dependences of $n$ and $k$ are specified in the form of a second- or third-degree polynomial in frequency. There is no need in this case to measure the $\phi(f)$ spectrum. The lines in Fig. 3.6 show the fit by $n = 3.59$; $k = 0.003$ for GaAs and $n = 14.45 + 3\cdot10^{-4}f$ $k = 0.29 - 4\cdot10^{-4}f + 5\cdot10^{-7}f^2$ for BaNaN ($f$ in GHz).

Most transparent materials reveal practically constant $n$ and $k$ linearly dependent on frequency. An example of this sort is demonstrated by materials commonly used in the far-infrared technique (Fig. 3.7).

Measurements "via oscillations" are the fastest and most convenient; they are completely automated in the "Epsilon" spectrometer and are brought up

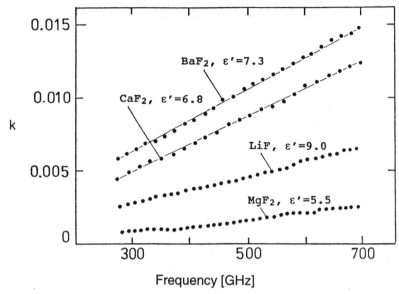

**Fig. 3.7.** SBMM absorption spectra of the simplest type: small values of $k \propto f$, $n \approx$ const. Points correspond on frequency to maxima of the interference pattern

to the level of mass routine measurements. Obtaining the data set shown in Fig. 3.7 takes 10–15 min, so dozens of materials can be characterized in a similar manner on the BWO spectrometer during one working day.

Having pointed out the simplicity and reliability of the registration of spectra on the BWO spectrometer, we ought to note that permanent control of the standing waves in the measurement path is necessary. What actually happens is that practically any SBMM spectrum inevitably contains distortions due to the standing waves. It is another matter that they can be taken into account by computer processing or averaged just during the experiment. The latter case is presented in Fig. 3.8.

### 3.1.3 Elements of Submillimeter Quasi-optics

**a) Detectors.** The BWO's radiation can be detected by both microwave methods – crystal detectors with point contact (detector heads, video-detectors), and methods of far-infrared spectroscopy – thermal and photo-electric detectors. All such methods have been described many times in the literature together with their characteristics [3.4, 5]. The usefulness of each detector is estimated according to several parameters: the working frequency range, sensitivity, rapidity, and operation conditions.

Detector heads are more suitable for the MM range, where they have an appropriate sensitivity in the video regime $\approx 10^{-11}$ $WHz^{-1/2}$ and a response time $\approx 10^{-9}$ s. They work at room temperature, are simple to operate, and are

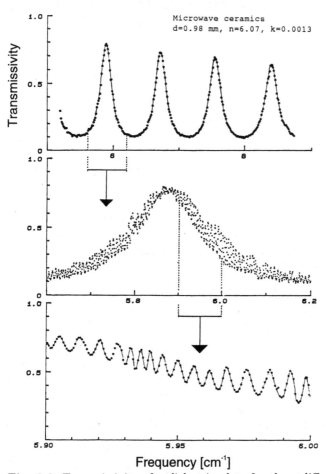

**Fig. 3.8.** Transmissivity of a dielectric plate for three different frequency scales. Resolution of the bottom spectrum is $0.001\,\mathrm{cm}^{-1}$

commercially available. Their main drawback is their frequency selectivity. The receiving unit of the detector head is always characterized by a complicated, frequency-cut apparatus function. The detector heads are normally used in a waveguide variant under the minimal frequency tuning conditions.

Among the photoelectric detectors based on the photo-effect, only a few are suitable for the SBMM part of the spectrum, the best of which is a cooled detector based on indium antimonide (n-InSb). This detector is wide-band, sensitive ($\approx 10^{-13}\,\mathrm{WHz}^{-1/2}$), has a response time of $\tau \approx 10^{-4}\,\mathrm{s}$, and as little as $10^{-7}\,\mathrm{s}$ in a magnetic field.

The thermal detectors, which work effectively at SBMM wavelengths, are germanium, carbon and superconducting bolometers, as well as pyroelectric and optical-acoustical (Golay cell) detectors. The bolometers are normally cooled, have high sensitivity $(10^{-11}$–$10^{-13}\,\text{WHz}^{-1/2})$, but comparatively large inertia $(t \approx 10^{-2}$–$10^{-3}\,\text{s})$.

The sensitivity of the Golay cells is about $10^{-10}$–$10^{-11}\,\text{WHz}^{-1/2}$, and that of the pyroelectrical detectors is several times lower. The main merits of these detectors are their ability to work at room temperature and their frequency-independent response in the whole SBMM wavelength range. These properties were decisive in choosing the Golay cell as the basic detector for BWO spectrometers. The value of the sensitivity has turned out to be of little importance while working with the intensive BWO radiation. We use the industrial Golay cells with a 0.5 mm thick and 6 mm in diameter polyethylene windows, produced in Russia and slightly modified by us. They are convenient and reliable devices and their properties do not change over many years. Their dynamical range (the linear part of the dependence of the response on the power of radiation) is 40 dB, and the saturation power is about $10^{-1}\,\text{mW}$. The Golay detectors most optimally correspond to the power characteristics of BWOs and measurement techniques developed on the basis of BWOs, working with radiation intensity overfalls 4–6 orders of magnitudes. The great majority of our measurements, including those described in the present review, are performed using the Golay cells.

In one type of our BWO-measurements a semiconductor room-temperature bolometer is used instead of the Golay cell. The main reason for this is its small dimensions and insensitivity to jolting. Based on the bolometer we have developed a scanner, able to register the distribution of the electromagnetic field in the plane perpendicular to the $k$-vector of the radiation. The bolometer is fixed on the edge of a 150 mm long pivot and can be moved step by step in horizontal and vertical directions by a mechanical scheme controlled by a computer. The working aperture of the scanner is $70 \times 70\,\text{mm}^2$, and the spatial resolution is of the order of the wavelength $\lambda$ (i.e., $\approx 1\,\text{mm}$ on average).

Figure 3.9 presents several examples of measurement done with this scanner. It is seen that the area of application of this device in combination with the BWO spectrometer is quite broad: contactless control of impurities, imperfections, and stress distribution in optically nontransparent dielectrics (introscopy), investigation of free carriers density in semiconductors, test and measurement of the thicknesses, registration of the fields, distribution in the quasi-optical beams, characterization of various quasi-optical devices – lenses, diaphragms, horns, etc.

b) Focusers (Lenses and Parabolic Mirrors). To form the quasi-optical beams we primarily use dielectric lenses, designed and manufactured according to the optical formula for a lens [3.13].

$$\frac{1}{F} = (n-1)\frac{2}{r} + \frac{d(n-1)^2}{nr^2},$$

where $F$ is the focal length, $r$ is the radius of curvature, $n$ is the refractive index of material, and $d$ is the lens thickness. The roughness of the surface

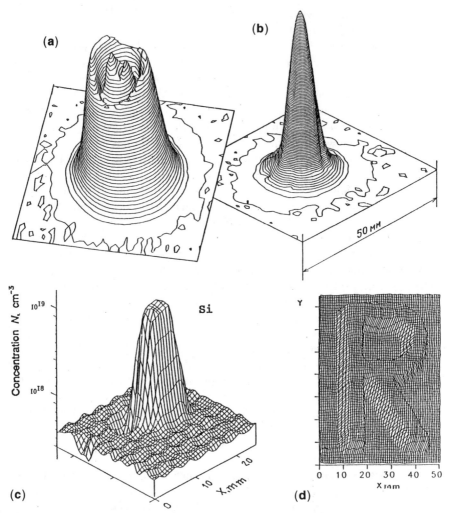

**Fig. 3.9 a–d.** Measurements on the scanner: (**a**) and (**b**) – radiation field distribution in the SBMM beam behind lenses for two different adjustments; (**c**) – density of free carrier distribution in an industrial silicon plate; (**d**) – contactless determination of the thickness ($\lambda$–1.5 mm, "phase portrait" of the paper letter "R")

must be much less than the radiation wavelength. In the wavelength range $\lambda \approx 3$–0.3 mm, this requirement can easily be fulfilled with standard turning without additional optical processing. Among the various materials transparent for the MM-SBMM radiation we prefer Teflon and polyethylene for lens fabrication. These are the most optimal from the viewpoint of both mechanical processing and electrodynamical characteristics. Teflon lenses are turned on a digital lathe and polyethylene ones are pressed in metal forms turned on a lathe. The standard parameters of our lenses are the following: the clear aperture is 50 mm, the focal lengths are $F = 50, 60, 80, 100,$ and 120 mm, and the typical thicknesses in the center are 10–20 mm.

At SBMM wavelengths the radiation losses in the Teflon and polyethylene lenses are very small (Table 3.1). However, the advantage of the polyethylene lenses becomes noticeable in the short-wavelength part of the SBMM region: in practice, with several lenses in the quasioptical path, the use of polyethylene at $\lambda \approx 0.3$ mm increases the detector signal about one order of magnitude as compared to Teflon. Our experience has shown that these lenses form quite a good instrumental beam. Their radiation patterns registered by the scanner reveal circular symmetry, negligible sidelobes (Fig. 3.9b), and a wide bandwidth.

Unfortunately, although they are very convenient and easily adjustable, dielectric lenses are unusable in reflectivity measurement schemes due to noticeable backward reflections to the quasioptical path and pronounced resonance properties. In these cases they are replaced by off-axis metallic parabolic mirrors. Such mirrors are cut from a big circular parabolic mirror, turned on a lathe from a massive duralumin piece. Our standard mirrors are rectangular slabs having a polished parabolic working surface of $60 \times 100$ mm$^2$ area with 100 and 200 mm focal lengths.

**Table 3.1 a,b.** Transparent materials for SBMM quasioptics: (a) – polymers, (b) – solids. $\lambda = 1$ mm ($f = 300$ GHz). Properties are partly sample-dependent

a)

| Material | n | k | $\varepsilon'$ | $\varepsilon''$ | R |
|---|---|---|---|---|---|
| Polyethylene | 1.41 | 0.0006 | 1.99 | 0.0017 | 0.03 |
| Teflon | 1.44 | 0.0015 | 2.07 | 0.0043 | 0.03 |
| TPX | 1.48 | 0.0020 | 2.19 | 0.006 | 0.04 |
| Paraffin | 1.5 | 0.0010 | 2.25 | 0.003 | 0.04 |
| Polystyrene | 1.5 | 0.003 | 2.25 | 0.009 | 0.04 |
| Plexiglass | 1.6 | 0.010 | 2.56 | 0.032 | 0.05 |
| Epoxide resin | 1.6 | 0.020 | 2.56 | 0.064 | 0.05 |
| Mylar | 1.8 | 0.015 | 3.24 | 0.054 | 0.08 |

b)

| Material | n | k | $\varepsilon'$ | $\varepsilon''$ | R |
|---|---|---|---|---|---|
| Ge | 3.99 | 0.02 | 15.9 | 0.16 | 0.36 |
| GaAs | 3.59 | 0.003 | 12.9 | 0.02 | 0.32 |
| Si | 3.43 | 0.004 | 11.8 | 0.03 | 0.30 |
| $Al_2O_3$ – ceramic | 3.15 | 0.002 | 9.9 | 0.013 | 0.27 |
| Mica | 2.50 | 0.004 | 6.25 | 0.02 | 0.18 |
| CV – diamond | 2.40 | 0.0005 | 5.76 | 0.002 | 0.17 |
| BN – ceramic | 2.14 | 0.001 | 4.58 | 0.004 | 0.13 |
| $SiO_2$ – glass | 1.96 | 0.001 | 3.84 | 0.004 | 0.10 |
| $SiO_2$ – ceramic | 1.76 | 0.003 | 3.09 | 0.01 | 0.08 |

**c) Wire Grids and Meshes.** Fine-structure metallic wire grids and cell meshes are the record-holders in MM and SBMM optics in the variety of their applications. They work in different schemes as radiation polarizers, beam splitters, couplers, semitransparent mirrors, filters, phase shifters, etc. They are attractive theoretically, as well, allowing for precise mathematical description of their electrodynamical properties. A large number of investigations have been devoted to the study of the electrodynamics of such systems [3.15–18].

Figure 3.10 shows the types of structures whose use we have mastered. Wire grids (a) are used mainly as polarization sensitive elements. They consist of an array of parallel tungsten wires affixed to mounting metal rings. Closely spaced wires reflect the electric field component $E_\parallel$ parallel to the direction of wires, and transmit the component $E_\perp$ perpendicular to the wires. The electrodynamics of the grids is determined by their geometrical parameters – wires diameter $D$, winding spacing (period) $L$, and also by the conductivity $\sigma$ of the metal and the ratio $æ = L/\lambda$. The parameter $s = D/L$ is called the filling coefficient of the grid. The tungsten grids which we fabricate and use in our BWO spectrometer have the following parameters:

clear aperture – 40 and 90 mm;

wire diameter – 8, 10, 15, 20, 25 μm;

wire spacing – from 30 to 500 μm with a step of 10 μm.

The specific electrodynamic characteristics of the grids are their transmission coefficients $T_\parallel$ and $T_\perp$ for linearly polarized waves $E_\parallel$ and $E_\perp$ at normal incidence to the plane of the grid. The transition from $T_\parallel$ to $T_\perp$ while rotating the grid by an angle $\alpha$ in its plane around the incident beam direction is governed by the simple relation:

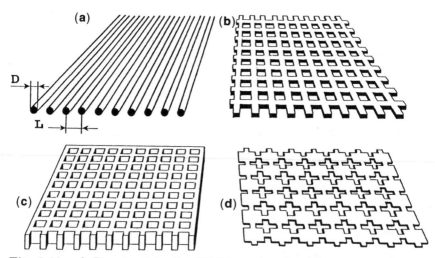

**Fig. 3.10 a–d.** Fine-structures for SBMM quasi-optics: (**a**) – wire grid ($D$ and $L$ are the diameter of wires and the winding period); (**b**) and (**d**) – electroformed thin metallic meshes; (**c**) – thick metallic plate with through holes

$T \propto \cos^2 \alpha$ .        $(T = T_{\parallel}$ at $\alpha = 0,$    $T = T_{\perp}$ at $\alpha = 90°)$ .

The same law also governs the corresponding reflection coefficients $R_{\parallel}$ and $R_{\perp}$, with a phase shift of $90°$. In this manner the rotation of the grid enables smooth redistribution of the energy between the reflected and transmitted waves. This redistribution process is accompanied by a smooth and simultaneous rotation of the planes of polarization in the transmitted and reflected beams. The energy absorption coefficient $A$ completes the energy balance:

$T + R + A = 1.$

Typically the losses $A$ of the metallic grids do not exceed $1\,\%$.

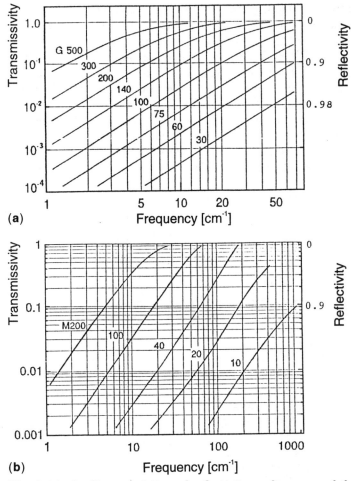

**Fig. 3.11 a,b.** Transmissivity and reflectivity $\nu s$ frequency of the fine structures: (a) – wire grids (G), field polarization is parallel to the wires; (b) – electroformed meshes (M), unpolarized radiation. Theory and experiment coincide within the graphical accuracy. Digits denote fine structure periodicity

We have studied the electrodynamics of the wire grids by the detailed BWO measurements [3.19] with the main conclusion that their properties can be perfectly described by the modern theoretical models [3.15–17]. As for the grids with æ < 0.3, their properties allow for the most simple analytic description [3.15]. For the longitudinal polarization ($E$ vector parallel to the wires), in particular, the complex transmission and reflection coefficients are:

$$\left\{ \begin{array}{c} T_E^* \\ R_E^* \end{array} \right\} = -\frac{1}{2}\left( \frac{1+iql_0}{1-iql_0} \mp \frac{1-iql_2}{1+iql_2} \right),$$

where $q = 2\pi/\lambda$ and the numerical coefficients $l_0$ and $l_2$ depend on the geometry and conductivity of the wires. Fig. 3.11a shows the transmissivity $T_\parallel$ for metallic wire grids made of conductors of circular cross section. The $T_\perp$ value is indistinguishable from unity on the scale presented.

Figure 3.12 shows a quasi-optical device based on the grids, designed by us for mutual transducing of linear, circular, and elliptic polarization of the SBMM radiation. It consists of a mobile metallic mirror controlled by a micrometer screw and a fixed wire grid placed in front of it. The radiation incident at an angle of 45° is divided into two waves with the linear and mutually orthogonal polarization. One wave is reflected by the wire grid, while the other is reflected by the metallic mirror. The phase shift between the two waves is determined by the separation between the mirror and the grid. Changing this distance one can obtain any type of polarization desired at the output of the transformer. Example of use of the tranducer is shown in Fig. 3.4c.

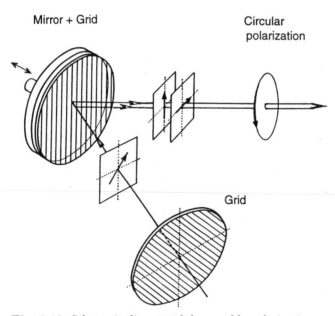

**Fig. 3.12.** Schematic diagram of the tunable polarization transformer

Meshes (Fig. 3.10 b–d) are much more complicated and diverse structures. They are fabricated by electroforming and also by many other methods, such as mechanical (drill), electrical discharge, electrochemical (etching), laser, etc. The electroformed meshes are thin metallic films (5–10 μm thick) of up to $150 \times 150\,mm^2$ area with fine clear cells of various configurations. Along with the wire grids, they have found the broadest range of application in our practice as:

1) mirrors of the Fabry-Perot resonators (square $a \times a$ cells, where $a = 40, 80, 100\ldots 260\,\mu m$);
2) polarization insensitive beam splitters (rectangular cells drawn out along the radiation traveling direction);
3) fine structure polarizers of large area (strips, without transverse links);
4) band pass filters.

**d) Fabry–Perot Interferometer.** Two semi-transparent mirrors placed face-to-face form the Fabry–Perot-interferometer (FPI). If the planes of the mirrors are adjusted to be parallel with a high enough accuracy, then the radiation falling normally on the system is trapped between the mirrors traveling repeatedly back and forth. The field energy partly penetrates through, is partly absorbed by the mirrors and partly scattered into open space beyond the mirrors' planes due to diffraction. The property of the system which is most commonly applied is its high transparency for radiation of wavelength $\lambda$ whenever the mirrors are separated by an integer number of half-wavelengths. Otherwise the FPI completely reflects back the intensity as soon as this condition is violated. As a result, the transmissivity (reflectivity) reveals very sharp peaks (gaps) when either the radiation frequency is swept, or, if the frequency is held constant, when the distance between the mirrors is changed.

The FPI is one of the most famous devices in modern optics, quantum radiophysics, and laser technology. By the 1980s it was mastered in full measure by the MM-SBMM quasioptics as well [3.20]. The main functions of the FPI at the MM-SBMM wavelengths is the frequency measurement, radiation filtering, dielectric measurements, gas spectroscopy, and plasma diagnostics. Irrespective of the kind of job, the measurable quantities are, as always, resonance frequency shifts, changes in the periodicity of the interference pattern, and the widths and amplitudes of the resonance maxima (minima).

Compared to optics, the requirements of mechanical accuracy for SBMM FPI are much lower and this makes these devices comparatively unpretentious, easily handled, and convenient to use. In particular, in our BWO-spectrometers we successfully use a very compact and simple construction without any mirror adjustment mechanism. Mirrors are just installed without additional manipulations into previously aligned nests having magnetic holders. Relative translation of the mirrors is realized with a standard micrometer screw.

Mirrors are wire grids or metallic meshes described in the previous section. Fastening rings for the mirrors have 40 mm clear apertures. They are made of

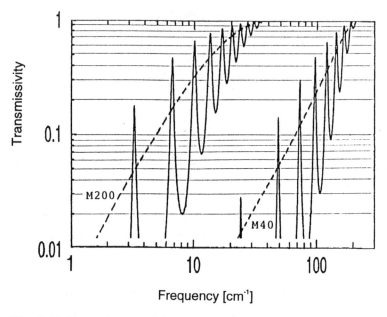

**Fig. 3.13.** Transmissivity of the Fabry-Perot interferometer (*solid*) and of the separate meshes of which it is composed (*dashed*) vs frequency. Theory and experiment coincide within the graphical accuracy

magnetic material, treated optically, and have a flatness and plane-parallelity not worse than ±2 μm over the entire diameter. This makes them changeable without further adjustment.

As applied to the radiation wavelength measurement, the FPI described provides current precision of 0.1 %. Figure 3.13 shows comparative transmissivity of separate mesh mirrors and the corresponding FPI. This figure helps to choose the proper mirrors for the given operating frequency.

**e) Thin Metallic Films.** Thin metallic films ($d \ll \lambda$, $\delta$; with $\delta$ – the skin depth) are of interest for the MM-SBMM quasi-optics, first of all, due to their technological characteristics: firstly, they can be prepared having a big area and, secondly, any needed thickness can be realized to provide the electrodynamic parameters $T$, $R$, $A$, convenient for work.

At first glance, the semitransparent metal films seem to be equivalent to the fine metal meshes and may be used along with them as beam splitters, mirrors, filters, couplers, etc. However, this is not so because of the large ohmic losses principally inherent in the semitransparent conducting films. If one attempts, for example, to use the metal film to split the incoming radiation into two equivalent beams ($T \approx R$), then 50 % of the energy is lost to heating of the film [3.21]. While this property results in poor quality mirrors and beam splitters, it makes the films appropriate as absorbing elements of the radiation detectors.

We have developed convenient quasi-optical film attenuators for our BWO-spectrometers, providing several discrete attenuation levels: $T = 30$, 10, 3, 1 % (5, 10, 15, 20 dB). The attenuator is a thin conducting layer deposited on a 5 μm thick Mylar film stretched over a circular metallic frame of 60 mm clear aperture. As a unit, the device is a block of four films changed by folding. The main merit of these attenuators is the complete absence of frequency dispersion of the attenuation coefficient in the whole MM-SBMM range. The absorbing film introduced into the quasioptical path serves also as a good discoupler, suppressing the standing waves and improving the quality of the spectra.

The transparency of the films allows for the investigation of the dielectric (electric) properties of the material of which they are made. Fig 3.14 shows transmissivity of the two superconducting NbN films, evaporated on both sides of a sapphire substrate [3.22]. The system is essentially a Fabry–Perot interferometer formed by two flat semitransparent mirrors (NbN films) and filled with a transparent dielectric (sapphire). It is extremely sensitive to the NbN conductivity: the intensities of the central maxima of the interference pattern increase by more than an order of magnitude in the superconducting phase. A strong decrease of the linewidth is also observed. The electrodynamic properties of the system are entirely described by the optical formulas of

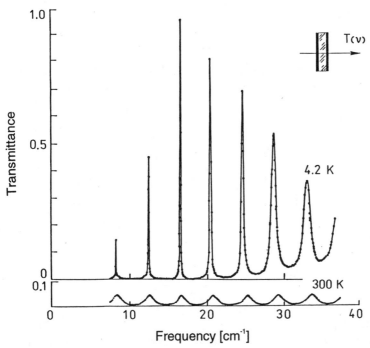

**Fig. 3.14.** Frequency dependence of transmittance of the NbN Fabry–Perot resonator (*the upper insert*) at room and liquid helium temperatures. Sapphire substrate thickness is 0.39 mm, NbN films thickness is 540 Å

a plane multilayer media [3.13], and this allows us to reliably evaluate the conductivity of the film material (NbN). The corresponding data is presented in the Sect. 3.2.9 devoted to electronic conductors.

A feature of BWO-measurements is an ability to observe the temperature evolution of the interference pattern in all details. It is illustrated in Fig. 3.15 which presents the transmissivity of a double-sided FPI formed by two superconducting YBaCuO films on a sapphire substrate. The superconducting phase transition in $YBa_2Cu_3O_{7-x}$ at 85 K decreases the level of the transmissivity at low temperatures, essentially increases the $Q$-factor of the interference maxima, and, what is the most striking, reverses the phase of the oscillations. Clearly seen is an almost complete disappearance of these oscillations in the spectrum at 80 K. This happens due to the impedance matching of the air-sapphire interface by a conducting $YBa_2Cu_3O_{7-x}$ film.

The phenomenon is similar to the one used in optics for an interferometric antireflection coating with a dielectric layer, where the first, second, etc., interference orders are used. In this context the matching by a thin metallic film can be regarded as an antireflection coating working in the zeroth interference order.

The matching condition results directly from the general multilayer optical formulae [3.13]. Neglecting the losses in sapphire one obtains

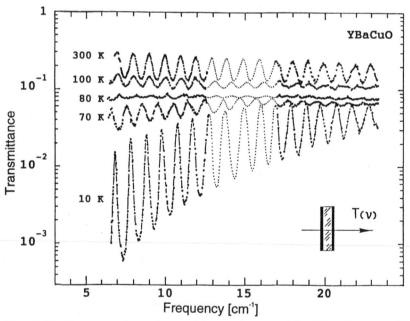

**Fig. 3.15.** Frequency dependence of transmittance of the $YBa_2Cu_3O_{7-x}$ Fabry–Perot resonator (*the bottom insert*) at different temperatures. $LaAlO_3$ substrate is 1.02 mm thick, film thickness is 150 Å . Note the matching phenomenon at 80 K and the reverse of the oscillating pattern phase at lower temperatures

$$d_0 = (n_S - 1)(120\pi\sigma_0)^{-1},$$

where $\sigma_0$ is the film conductivity (in $cm^{-1}$), $d_0$ is the film thickness, $n_S$ is the refractive index of the substrate. A feature of the relationship is the absence of the radiation frequency, i.e., the absolute frequency independence of the matching effect. It may be violated (the modulation can appear in the transmission spectrum) due to dispersion of the optical properties of the materials ($n_S$ or $\sigma_0$), but at low frequencies, including our experimental SBMM frequency range, the dispersionless behavior of $n_S$ and $\sigma_0$ is the most common case, when we deal with a dielectric having low losses and a film of normal metal governed by the Drude law for the frequency behavior of the conductivity ($\sigma_0 = \sigma_{SBMM} = $ const). So, in fact, the metallic matching at SBMM wavelengths reveals an extremely wide bandwidth.

f) **Plane Dielectric Slab.** All our techniques of dielectric measurements on the BWO-spectrometers are based on the interaction of a plane electromagnetic wave with a plane layer of substance. The examples already described are the measurements employing a dielectric plate (Fig. 3.5, 6) and conductivity measurement of superconducting films (Figs. 3.14, 15). In general, one can consider the interaction of the electromagnetic field with a multilayer system consisting of an arbitrary number of layers. The clear merit of this system, as an object under study, is the simplicity of its fabrication and the possibility of rigorous mathematical treatment according to the Fresnel formulas.

An important information channel is provided by the interference of the waves in a layered system. Infrared spectroscopists of past years did not like this phenomenon and did their best to avoid it in the measurements. A well-known way of doing that is to make the sample wedge-shaped. On the contrary, the interference is helpful and desired in our measurements since it allows for computer simulations using the rigorous formulae. Its function is to a) increase the efficiency of the interaction of the radiation with the substance (interaction time and path) and by that increase the accuracy of the amplitude measurements and b) provide an opportunity to calculate the phases $\phi$ and $\psi$ in addition to the amplitude characteristics $T$ and $R$. This is especially important during reflectivity measurements.

Interference of waves inside the high-quality dielectric resonator (tester) serves as the basis of our method of measurement of reflectivities of nontransparent materials [3.23]. The $R$ and $\psi$ values are extracted from the comparison of two interference patterns: the reflectivity spectrum of the tester alone, and of the same tester with the absorbing sample under study, in contact with its rare face. The needed information on $R$ and $\psi$ is contained in the frequency shift of the resonance reflectivity minima and in the change of their depth. We have recently succeeded in application of the tester method to the highly conducting crystal $K_{0.3}MoO_3$ (blue bronze) [3.24].

The computer modeling of the penetration of the electromagnetic waves through the multilayer structure is embedded into the software of our BWO-spectrometers and may be used just during the measurement process. In addition, it strongly simplifies the solution of many technological problems. Using the real-time computer simulation makes it possible to take into account the influence of the paste layer while gluing the sample on a substrate, of cryostats and thermostats stratified windows, cuvette walls, damaged surface layers of the samples (after the polishing), etc.

In closing this section we present the plane dielectric layer in one more capacity – as a frequency meter of the monochromatic radiation ($f$-meter). In this case we have done our best to avoid the interference in the sample and to realize a purely exponential frequency behavior of the transmissivity. We have chosen a material with a high enough absorptivity, frequency-independent $k$, and low enough reflectivity ($R < 10\%$), and gotten a simple linear function $T(f)$ (in the logarithmic scale), uniquely connecting $T$ and $f$. As a result, a calibrated dielectric plate introduced into a radiation beam indicates insantly the value of the frequency $f$ *via* $T$. The accuracy of such a $f$-meter is not very high, $\approx \pm 2\%$. Its advantage is its extreme simplicity, ease of handling, and ability to determine the frequency in a pulsed regime.

**g) Materials at MM-SBMM Waves.** Irradiation of various samples on a BWO-spectrometer reveals pellucidity for SBMM radiation of a great variety of optically opaque and transparent materials – paper and carton, cloth and wood, marble and concrete, butter and chocolate, oil products and coal, different radio- and optical materials. This first impression is sketched in Table 3.2. It is seen that water has the highest absorptivity among common materials. For this reason its presence in the samples (water content) can easily be detected. Obviously, the sensitivity of the BWO measurements to water can find practical applications.

In the amount of reference data on dielectric properties of materials, the SBMM range ranks much lower than neighboring MM and infrared regions. The BWO-spectrometers are highly promising for solving the presently urgent task of filling up this gap. The second part of the review is devoted essentially to this topic. Here we restrict ourselves to an example of room temperature data on some daily used materials of the BWO-technique. They are presented in Fig. 3.16a, b and in Table 3.1. The feature of the plots is their combinative character: they are formed by the merger of the SBMM (double shaded) and IR (lines) transmissivity data. The intermediate points (shaded) are obtained in the framework of the multioscillator dispersion model.

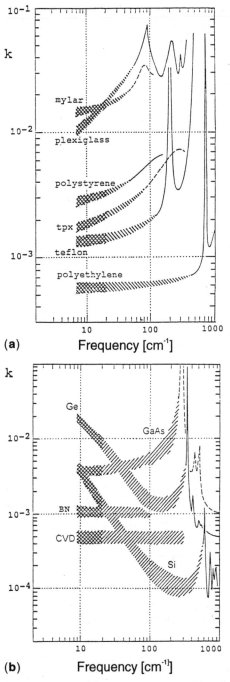

**(a)**     Frequency [cm⁻¹]

**(b)**     Frequency [cm⁻¹]

**Fig. 3.16 a,b.** Absorption spectra of low loss isotropic materials for SBMM-quasi-optics: (a) – polymers, (b) – solids (high resistivity semiconductors). Properties are partly sample-dependent (*shaded*). Room temperature

**Table 3.2.** Comparison data on dielectric properties of common materials at $f \approx 300\,\text{GHz}$ ($\lambda = 1\,\text{mm}$); $\alpha$ is absorption coefficient, $d_{1/3}$ is thickness of 30 % transmissivity

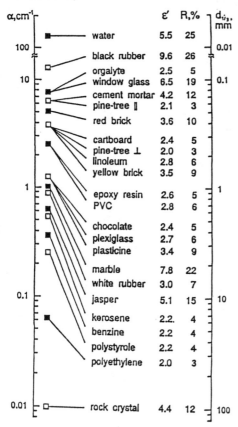

| | $\varepsilon'$ | R,% |
|---|---|---|
| water | 5.5 | 25 |
| black rubber | 9.6 | 26 |
| orgalyte | 2.5 | 5 |
| window glass | 6.5 | 19 |
| cement mortar | 4.2 | 12 |
| pine-tree ‖ | 2.1 | 3 |
| red brick | 3.6 | 10 |
| cartboard | 2.4 | 5 |
| pine-tree ⊥ | 2.0 | 3 |
| linoleum | 2.8 | 6 |
| yellow brick | 3.5 | 9 |
| epoxy resin | 2.6 | 5 |
| PVC | 2.8 | 6 |
| chocolate | 2.4 | 5 |
| plexiglass | 2.7 | 6 |
| plasticine | 3.4 | 9 |
| marble | 7.8 | 22 |
| white rubber | 3.0 | 7 |
| jasper | 5.1 | 15 |
| kerosene | 2.2. | 4 |
| benzine | 2.2 | 4 |
| polystyrole | 2.2 | 4 |
| polyethylene | 2.0 | 3 |
| rock crystal | 4.4 | 12 |

## 3.2 Dielectric BWO-Spectroscopy of Solids

The opportunity to perform fast measurements of the dielectric response function by the BWO-spectrometer allowed us to observe its frequency and temperature properties in a variety of different manifestations, in materials of different classes.

Via SBMM measurements we have established the direct connection between the two independent schools in radio spectroscopy: high-frequency – Infrared and Raman spectroscopy dealing with the lattice vibrations and low-frequency dielectric measurements dealing with relaxations and conductivity. The relative position of the schools on the frequency axis is shown in Fig. 3.17. The above mentioned microscopic mechanisms are shown schematically by solid, dashed, and dotted lines. Typically, they are strongly smeared over a huge frequency interval and overlapped. This is the reason for a persis-

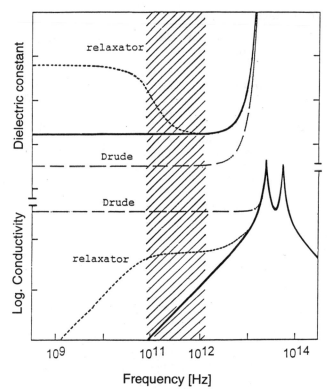

**Fig. 3.17.** Three strongest absorption mechanisms in solids: lattice resonances, dielectric relaxation, and conductivity (Drude mechanism). *Shaded* is the SBMM wavelength range, which merges two qualitatively different shapes of the dielectric response

tent tendency in the spectroscopy of recent years to extend the observation windows and to build up the panoramas. Dielectric response is a uniquely wide-band measurable function: approximately 20 spectral decades from $10^{-5}$ to $10^{15}$ Hz are accessible today for experimentalists. Its general properties directly result from the first principles – the energy and impulse conservation laws, and the causality principle [3.25–27].

We describe below the dielectric response function behavior that we have observed in the range $10^{10}$–$10^{12}$ Hz with the help of BWO spectrometers [3.10–12, 28]. Table 3.3 presents a list of crystals which have been investigated. They are divided roughly into several groups according to the similarity of the observed phenomena. Below we present the most characteristic examples.

### 3.2.1 Dielectric Spectra of Simple Dielectrics

We arbitrarily call those materials simple dielectrics whose dielectric response obeys the rules of classical phonon physics: occurrence of sharp peaks in the

**Table 3.3.** Materials which have been studied

| Ferroelectrics | Incommensurate crystals | Semiconductors | Superionics |
|---|---|---|---|
| $TiO_2$, $SrTiO_3$, $BaTiO_3$, $KTaO_3$, PLZT, $Ag(Nb,Ta)O_3$, $Pb_5Ge_3O_{11}$, $Sn_2P_2S_6$, $Ba_2NaNb_5O_{15}$, $KTiOPO_4$, *Family:* KDP, DKDP, RDP, KDA, ADP, (R,A)DP, D(R,A)DP *Rochelle salt family:* RS, DRS, ARS, (TU)RS, LTT TGS, DTGS, $RbHPO_4$, $RbHSO_4$, $RbDSO_4$, $NaNO_2$, $NaNO_3$, KSCN, $(NH_4)_2SO_4$, $LiNH_4SO_4$, $KH_3(SeO_4)_2$, $KD_3(SeO_4)_2$, $Gd_2(MoO_4)_3$, TSCC, $Li_2Ge_7O_{15}$, betains BA, BP | $K_2SeO_4$, $Rb_2CoCl_4$, $Rb_2ZnCl_4$, $Rb_2ZnBr_4$, $K_2ZnCl_4$, $(NH_4)_2BeF_4$, $Sr_2Nb_2O_7$, $BaMnF_4$, $CsCuCl_3$, $SC(NH_2)_2$, betain BCCD | Si, Ge, GaAs, InP, Se, InSe, $ZnGeP_2$, $Te_2Br$, PbI, $VO_2$, SbSI, $TlSbS_2$, $UPt_3$, *Family:* $TlGaSe_2$, $TlInS_2$, $TlInSe_2$, $TlGaS_2$, $TlGa(Se,S)_2$ | AgI, $RbAg_4I_5$, $AgI\text{-}Ag_2WO_4$, $Na\text{-}\beta\text{-}Al_2O_3$, $Na\text{-}\beta''\text{-}Al_2O_3$, $Na_3Sc_2(PO_4)_3$, $Cs_3H(PO_4)_3$, $RbCu_4Cl_3I_2$ |

| Dielectrics | Magnets | Low-dimensional conductors | Super-conductors |
|---|---|---|---|
| $SiO_2$, $GeO_2$, $Al_2O_3$, $(HfO_2, ZrO_2) + Y_2O_3$ (fianits), polymers, MW ceramics, IR materials, copolymers (VDF-FTE), KBr, LiF, NaCl, $CaF_2$, $CaCO_3$, $MgF_2$, CsI, MgO, $GdAlO_3$, $LaAlO_3$, $LaGaO_3$, $NdGaO_3$, $BaLaGa_3O_7$, $SrLaGa_3O_7$, $SrLaAlO_4$, $CaNdAlO_4$ | $YFeO_3$, $TmFeO_3$, $DyFeO_3$, $HoFeO_3$, $SmFeO_3$, $SmTbFeO_3$, $\alpha\text{-}Fe_2O_3$, $TbCrO_3$ | TTF-TCNQ, $MEM(TCNQ)_2$, $MTPP(TCNQ)_2$, $MTPA(TCNQ)_2$, $\alpha\text{-}(BEDT\text{-}TTF)_2I_3$ $(BEDT\text{-}TTF)_2Cu(NCS)_2$, $(NbSe_4)_3I$, $(TaSe_4)_2I$, $1T\text{-}TaS_2$, $K_{0.3}MoO_3$ | $YBa_2Cu_3O_{7-\delta}$, $(La,Sr)_2CuO_4$, NbN, NbC |

absorptivity and the loops of dielectric permittivity in the infrared region and their absolute absence at smaller frequencies – at SBMM, MM and radio-wavelengths. The SBMM range just joins the lattice absorption region, but still does not contain any of the absorption peaks. The simple dielectrics are comparatively transparent here and have dielectric permittivities $\varepsilon'$ practically equal to the static value $\varepsilon_0$.

The main practical and scientific interest is associated with the study of the residual loss background and mechanisms of its origin. Among these mechanisms, the contribution of the infrared peak tails, multiphonon processes, weak relaxations, and conductivity from imperfections and impurities should be regarded as the most important.

Figure 3.18 shows the panorama of the absorption spectra of a few famous materials – quartz, sapphire, CVD-diamond, and $CaF_2$ single crystal.

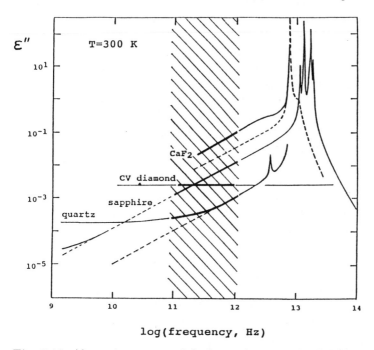

**Fig. 3.18.** Absorption spectra of the lowest-loss crystals. *Solid lines* – experiment, *thick segments* – BWO data, *broken lines* – one phonon contribution, calculated from the infrared reflection spectra. *Shaded* is the SBMM wavelength range

The panorama is obtained by merging our BWO and infrared data with the literature on radio-frequency data. We would like to demonstrate here the place which the SBMM fraction of the spectrum occupies relative to properties of real crystals and how different absorption mechanisms look in the panorama. In sapphire the SBMM losses are perfectly described by the tails of infrared phonons; in $CaF_2$ and many other simple ionic crystals the contribution of the multiphonon processes is clearly pronounced. Diamond is singled out by the complete absence of the single-phonon absorption peaks in the dielectric spectra. For this reason, its residual losses are very small and are practically frequency-independent.

### 3.2.2 Soft Modes in Ferroelectrics

All crystals famous for their extraordinary properties – ferroelectrics, superionics, dipole glasses, incommensurate crystals, superconductors, metal-dielectrics, etc., – gain these properties as a result of phase transitions. The existence of phase transitions in substances determines to a large extent the surrounding world, and that is why the problem of phase transitions is among the fundamental and the most hot problems of modern physics.

In the early 1960s the idea of the dynamic origin of the structural phase transitions in crystals was developed [3.29–31]. According to the theory, long before the transition into the new state takes place, there already exists in the crystal a temperature unstable lattice vibration that contains a new quality in it. It was predicted that as the temperature approaches the transition point the vibration should soften and its frequency should necessarily become zero at this point. This specific vibration was called the *soft mode*.

Soft modes in ferroelectrics are of special interest due to their activity in the infrared spectra and very clear manifestation in the frequency-temperature behavior of the dielectric response function. Usually, the ferroelectric soft modes occupy the lowest frequency position in the rank of the infrared lattice absorption peaks. Their remarkable property for the spectroscopists is that while the temperature is changed towards the phase transition, they separate from this group and move over the spectrum towards lower frequencies. They fill up the previously unoccupied spectral space at frequencies below $v \approx 10^{12}$ Hz.

Today, the science of ferroelectric soft modes gives a very nicely complete picture, presented in Fig. 3.19. The static temperature anomaly of the dielectric permittivity $\varepsilon_0(T)$ observed in ferroelectrics obeys the well-known Curie–Weiss law, discovered experimentally long ago:

$$\varepsilon_0(T) \propto (T - T_C)^{-1},$$

where $T_C$ is the Curie temperature.

Independent infrared measurements reveal the Cochran behavior of the frequencies of the soft modes (transverse modes):

$$\omega_t \propto (T - T_0)^{1/2},$$

where $T_0$ is the soft mode condensation temperature. The fundamental Lyddane–Sachs–Teller relation bridges these two phenomena [3.32]:

$$\frac{\varepsilon_\infty}{\varepsilon_0(T)} = \frac{\omega_t^2(T)}{\omega_l^2},$$

where $\omega_l$ and $\varepsilon_\infty$ are the frequency of the longitudinal vibration and the corresponding high-frequency dielectric permittivity, both essentially temperature-independent. Strictly speaking, the above relates to $\Delta\varepsilon = \varepsilon_0 - \varepsilon_\infty$, but $\varepsilon_\infty$ is considered to be negligible. It is seen how simply and rigorously the divergence of the static dielectric permittivity at the phase transition point (left graph), and the temperature evolution of the soft mode (bottom graph), are connected by a fundamental relation.

Leaving the infrared spectrum and moving toward the low frequencies, the soft modes leave the working range of the infrared spectrometers, which makes them difficult objects to be observed by conventional methods. By its frequency position, BWO spectroscopy seems to be especially intended for interception of the soft modes and their detailed study. We have performed this job on the crystals enumerated in the first two columns of Table 3.3. A

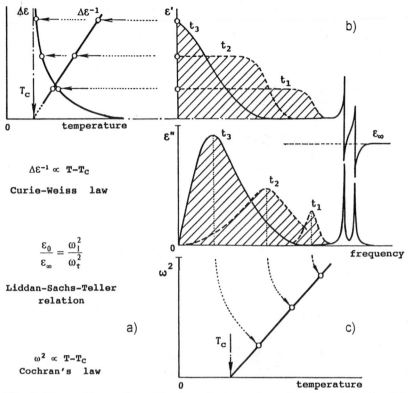

$\Delta\varepsilon^{-1} \propto T - T_c$

Curie-Weiss law

$$\frac{\varepsilon_0}{\varepsilon_\infty} = \frac{\omega_l^2}{\omega_t^2}$$

Liddan-Sachs-Teller relation

$\omega^2 \propto T - T_c$
Cochran's law

**Fig. 3.19 a–c.** Conception of the ferroelectric soft mode: (**a**) – soft mode dielectric contribution $\Delta\varepsilon$ and $\Delta\varepsilon^{-1}$ $vs$ temperature; (**b**) – ferroelectric soft mode (*shaded*) in the dielectric spectra at different temperatures $t_1$, $t_2$, $t_3$; (**c**) – soft mode's squared frequency $vs$ temperature. Note rigorous correlation between microscopic (**c**) and macroscopic (**a**) phenomena

great variety of the pre-transition phenomena have been observed [3.11, 33]. It turned out to be unexpectedly difficult to evaluate the data on the classical ferroelectrics – barium titanate (BaTiO$_3$), Rochelle Salt (RS), potassium dihydrophosphate (KDP), triglycine sulfite (TGS), and others, for which the theory of ferroelectricity has actually been developed. And *vice versa*, a more simple picture is presented, as a rule, by the new exotic ferroelectrics of complex composition – Tris Sarcosine Calcium Chloride (TSCC), benzil, and Betaine Calcium Chloride Dehydrate (BCCD).

We begin with a simple example. Fig. 3.20 shows the ferroelectric soft mode in Lithium Thallium Tartrate (LTT). It has a spectrally well at fixed temperature pronounced lonely absorption line with a Lorentz shape, and is extremely temperature-dependent. The data are described well at fixed temperature by the formula of a simple harmonic oscillator:

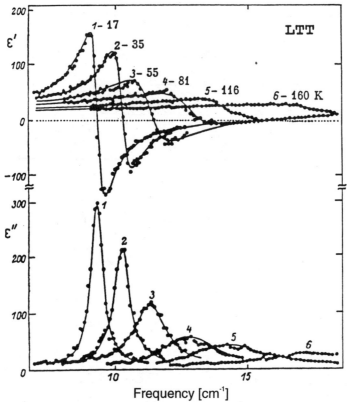

**Fig. 3.20.** Submillimeter dielectric spectra of LTT crystal. *Points* – experiment, *lines* – oscillator fit. Temperature-dependent absorption line is a ferroelectric soft mode

$$\varepsilon(\omega) = \frac{S}{\omega_0^2 - \omega^2 + i\omega\gamma}, \tag{3.5}$$

where $S$, $\omega_0$ and $\gamma$ are the strength, frequency, and damping of the oscillator, respectively. The values found, for the whole set of temperatures form the temperature dependences. They are presented in Fig. 3.21. To the first approximation the picture agrees well with the soft mode conception presented in Fig. 3.19: a linear temperature behavior of the squared frequency, the hyperbolic divergence of the dielectric contribution, temperature independence of the oscillator strength are observed.

Using Fig 3.20 as a typical example of the BWO dielectric measurements, let us take note of their completeness and high information level. The spectra consist of pairs of $\varepsilon'$ and $\varepsilon''$ points measured independently (without exploitation of the integral Kramers–Kronig relations) with high accuracies and presented in absolute values. Such experimental material allows for di-

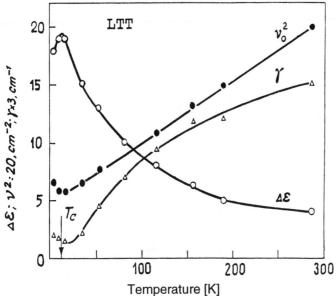

**Fig. 3.21.** Soft mode parameters $vs$ temperature in LTT crystal: $\nu_0$ – frequency, $\Delta\varepsilon$ – dielectric contribution, $\gamma$ – damping

rect quantitative comparisons with analogous data of other dielectric measurements (microwave and infrared) and with the theory as well [3.33].

Figure 3.22 presents example of ferroelectric soft modes in conventional ferroelectric – lead germanate (PGO). Similar pictures are also chacteristic for $BaTiO_3$, KDP and TGS. This new example does not differ in essence from the first of Fig. 3.20, however it is not so visual due to the much higher damping of the soft mode. And while the $\varepsilon'(\omega)$ and $\varepsilon''(\omega)$ spectra in PGO still obey the oscillator model with large damping ($\gamma > \omega_0$), the modes in TGS, $BaTiO_3$ and KDP are so damped that they correspond rather to a simple relaxation model:

$$\varepsilon(\omega) = \frac{S}{1 + i\omega\tau}, \tag{3.6}$$

where $S$ and $(2\pi\tau)^{-1}$ are the relaxator strength and the characteristic frequency, respectively. Normally such broad relaxation modes do not go into the operating range of the BWO-spectrometers and cannot be fully registered. Experimentally measurable $\varepsilon'(\omega)$ and $\varepsilon''(\omega)$ spectra grasp only a fraction of the dispersion. However, as a rule, this is enough to complete the picture within the model. Absolute values of $\varepsilon'$ and $\varepsilon''$ together with the slopes of the $\varepsilon'(\omega)$ and $\varepsilon''(\omega)$ curves, strictly and unambiguously determine the choice of $S$ and $(2\pi\tau)^{-1}$. The model parameters calculated for PGO are presented in Fig. 3.23. What is most important here is that the soft mode frequency does not vanish at the transition point as is expected from the soft mode

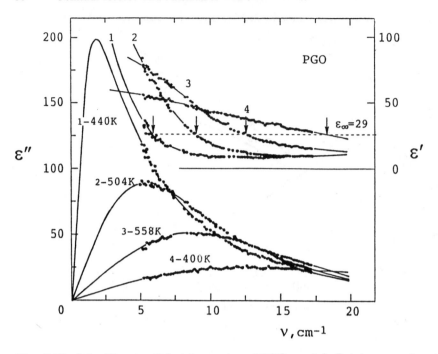

**Fig. 3.22.** Submillimeter dielectric spectra of PGO crystal. *Points* – experiment, *lines* – overdamped oscillator fits. Broad temperature-dependent absorption line is a ferroelectric soft mode. *Arrows* indicate the soft mode frequency

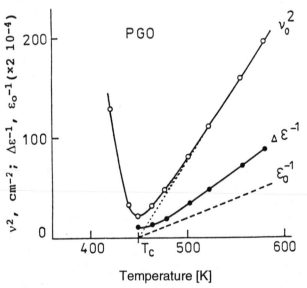

**Fig. 3.23.** Soft mode parameters $\nu s$ temperature in PGO crystal, calculated within the framework of oscillator model

conception: $\omega_0(T_C) \neq 0$. This topic relates to the problem of the central peak and is discussed in the next section.

Highly unexpected are the SBMM properties of the Rochelle Salt (RS) (Fig. 3.24). Already the visual appearance of the $\varepsilon'(\omega)$ and $\varepsilon''(\omega)$ spectra reveals a nontrivial fact: the phonons are directly involved in the phase transition mechanism (the narrow peak at $21\,\mathrm{cm}^{-1}$ at low temperatures). The temperature evolution of the soft mode evidently develops from this peak, i.e., as if the displacive mechanism lies in the basis of the phase transition. So far, RS is positively considered to belong to an alternative class of ferroelectrics, of the order-disorder type.

The temperature dependence of $(2\pi\tau)^{-1}$ in RS reveals a very abrupt increase with decreasing temperature, which is in serious contradiction to the usual linear law of $(2\pi\tau)^{-1}$ variation in relaxational ferroelectrics. We found that $(2\pi\tau)^{-1}$ in RS more likely obeys a cubic law:

$$(2\pi\tau)^{-1} = 1.07(T_0 - T)^3 \cdot 10^5 \mathrm{cm}^{-1},$$

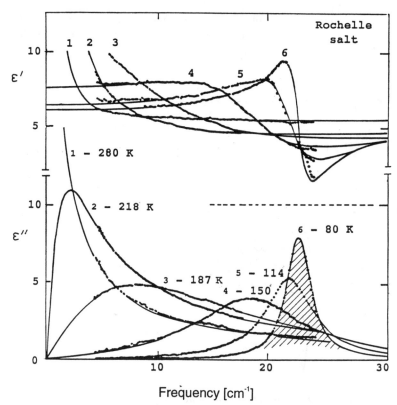

**Fig. 3.24.** Submillimeter dielectric spectra of the Rochelle salt. *Points* – experiment, *lines* – oscillator/relaxator fits. The soft mode behavior of spectra is developed from the lattice resonance (*shaded*)

where the temperature $T_0 = 276\,\mathrm{K}$ proved not to be the phase transition point at 255 K as would be expected, but rather a middle point of the ferroelectric phase.

Based on this discovery we have offered a way to substantially improve the thermodynamical description of the RS properties, considering the $T_0$ point as a *double critical point* [3.34]. From the analysis of the experimental data, we have calculated the coefficients of expansion of $A \propto \chi^{-1}$, where $\chi$ is the dielectric susceptibility in powers of the reduced temperature of the ferroelectric phase $t = (T - T_0)/T$, and found:

$$A = -0.0647 + 11.744t^2 - 14.814t^3 - 45.536t^4,$$

which differs principally from the conventional form $A \propto T - T_C$ by the absence of the linear term.

Using this $A$ in the thermodynamic potential, we have uniquely and consequently described the thermal, dielectric, optical, and elastic properties of RS related to the ferroelectric phase transition [3.35]. The most graphic result from this line is shown in Fig. 3.25. It clarifies the origin of the well-known exotic concentrational RS-ARS phase diagram (ARS is the ammonium RS) [3.36]: the latter is proved to be a cut of a two-body hyperboloid with a saddle point $X_0$ (ammonium content) $= 9\,\%$, $T_0 = -37°\mathrm{C}$, $P_0 = -2\,\mathrm{kbar}$.

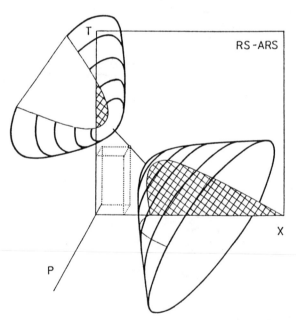

**Fig. 3.25.** Famous RS-ARS concentrational phase diagram of the Rochelle salt as a section of the two-body hyperboloid. Shaded is an area of the ferrophase

### 3.2.3 Central Peak

The simple, clear picture of the soft mode behavior is typically disturbed in the vicinity of the phase transition. This can easily be seen in Fig. 3.21 where the soft mode does not completely soften at the transition point. The soft mode contribution in LTT $\Delta\varepsilon \approx 20$ does not reach essentially the static value $\varepsilon_0 \approx 1000$.

The source of such disturbances is qualitatively understood at present [3.37]. During the temperature evolution, the soft mode inevitably interacts with the low-frequency degrees of freedom and it makes the shape of the dielectric response much more complicated. A single mode model becomes inadequate in this situation and has to be replaced with a more complicated multiparameter model. The first step in this direction is a model of two coupled oscillators [3.38]. Within this model both the weakening of the temperature variation of the soft mode frequency (the "soft mode" now is only one of the coupled oscillators, the higher frequency one), and the modesty of its dielectric strength are explained [3.33]. From the viewpoint of the physics of oscillations, a process of energy repumping from one oscillator to another occurs. It is obvious that this phenomenon may include more than two oscillators.

The SBMM part of the spectrum seems to be the first frontier where the soft mode meets, on its way to zero frequency, the low-frequency degrees of freedom which are nominally not active in the dielectric spectra. The soft mode pumps them, making them active. A question arises about the nature of the unknown excitations. Phenomenologically, they are always certain relaxations of different microscopic origin in different cases. The study of these low frequency relaxations pumped by the soft mode has only recently begun. It is done by dielectric measurements at frequencies below $10^{12}$ Hz, i.e., in the region where the excitation spectra of conventional dielectrics are normally empty. While interpreting the relaxations, such phenomena as cluster and domain dynamics, excitations of glassy and incommensurate phases, and ionic transport, are typically discussed. We cover some of these points in the following sections.

It is time now to come back to the title of the present section, the term *central peak*, which is widely present in the literature devoted to the low-frequency dynamics of ferroelectrics. It originates from neutron and Raman scattering experiments [3.29]. In contrast to dielectric spectroscopy, these methods are restricted in working frequencies by approximately 1–10 cm$^{-1}$, where the soft mode just begins to repump into the low-frequency relaxations. In the methods mentioned the power of low-frequency processes is released integrally in a very narrow frequency band close to the excitation line. In a normal scale (not logarithmic as in Fig. 3.17) this is seen as an ignition of an intensive scattering peak at zero frequency, where the name *central peak* originates from.

So, the central peak in neutron or Raman scattering is a good indication of possible low-frequency dynamics in the substance. Its spectral consistence, however, can be revealed only by other, lower frequency experimental techniques. In this respect, dielectric spectroscopy has a great advantage. Owing to an practically unlimited working frequency interval, one is able to study the low frequency dynamics independently within the framework of a single experimental method. The central peak does not make any sense in dielectric spectroscopy since there is no concept of zero frequency.

### 3.2.4 Dynamics of Incommensurate Phases

It was found in the 1960s, as a result of a thorough structural investigations, that besides the basic lattice periodicity, some dielectric crystals have additional structural modulations along certain directions [3.39]. Moreover, unlike crystals with superstructure, where the superstructure period is a multiple of the primary period and the crystal retains the translational symmetry along three dimensions, the fourth periodicity in the discovered systems develops independently, and is incommensurate with any of the basic ones. Incommensurate ferroelectrics are characterized by an incommensurate phase which precedes the ferroelectric phase upon cooling the crystal from normal to a polar phase. In this intermediate phase the structure exhibits a frozen-in polarization wave with a period incommensurate with, and much larger than, the lattice spacing. This period grows upon approaching the ferroelectric transition and the polarization wave changes into a periodic ferroelectric domain structure near the ferroelectric transition. From the point of view of lattice dynamics, the incommensurate phase enrichs by new low-frequency vibrations, namely by fluctuations of the amplitude and phase of the incommensurate modulation wave, the so-called *amplitudons* and *phasons* [3.40].

Phasons are the principal peculiarity of lattices with incommensurate modulations. By their outward appearance on the dispersion pattern, the phason branches are very similar to the acoustic ones. However, unlike the latter, whose damping factor is proportional to $q^2$ and vanishes at $q \to 0$, phasons have finite damping at $q \to 0$.

It was an intriguing challenge for experimentalists of the 1970s to prove the existence of phasons, and we have involved the BWO-spectroscopy in this process. Excitations of the phason type have been observed in a number of crystals with an incommensurate phase, such as theauria (TU), BCCD, and $Rb_2ZnCl_4$ [3.11], but the most vivid effect has been found in the classical incommensurate $K_2SeO_4$ crystal (potassium selenate). Being an improper ferroelectric, it is characterized by dynamics which is not disturbed by the dipole soft mode.

In Fig. 3.26 the manifestation of the phason in the SBMM spectra is presented. It gives a sharp absorptivity peak in the temperature dependences for the incommensurate phase (in the interval between $T_i = 129\,\mathrm{K}$

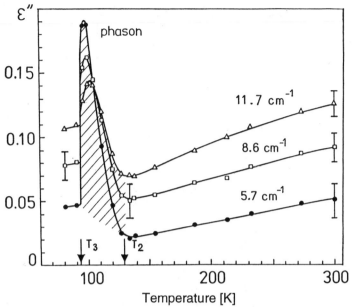

**Fig. 3.26.** Submillimeter absorptivity of $K_2SeO_4$ crystal $vs$ temperature

and $T_C = 93\,\text{K}$), and the spectrum of these losses leaves the SBMM range towards lower frequencies. The temperature behavior of the relaxation strength and the characteristic frequency of the discovered excitation, obtained from the SBMM data, agreed with concepts of the behavior of an inhomogeneous phason [3.40]. The full beauty of the phason in $K_2SeO_4$ was later registered in low-frequency dielectric measurements and clearly interpreted as a manifestation of the oscillating motion of the domain-like incommensurate wave of polarization [3.41].

### 3.2.5 Brillouin Zone Folding

In a number of crystals we have discovered a very striking phenomenon which lies in the fact that at low temperatures SBMM dielectric spectra split into a set of extremely narrow intensive lines [3.10, 11]. Examples of such behavior are presented in Fig. 3.27. It was found that the fine structure of the spectra originates from the multiplication of the unit cell accompanying structural phase transitions of a certain type. While the unit cell is multiplied $n$ times, an $n$-time folding of the Brillouin zone occurs, and the phonons from its interior, formerly not active in the infrared and Raman spectra, come to the $q = 0$ axis. Obviously, the folding involves both the optical and the acoustic branches. New absorption lines at the $q = 0$ axis originating from the optical branches fall on the higher frequency part of the spectrum and are accessible for registration by the conventional techniques of Raman and infrared spectroscopy. Deciphering of such spectra, however, is not a simple

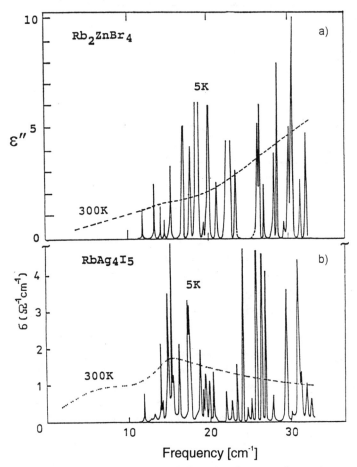

**Fig. 3.27 a,b.** Fine structure of the submillimeter absorption spectra in the crystals: (a) – rubidium zinc bromide; (b) – rubidium silver iodide

task since the comb of new resonancies is superimposed onto the spectrum already occupied by absorption peaks.

It is a feature of BWO-spectroscopy that owing to its low frequency working range ($v < 10^{12}$ Hz) it is able to trace the acoustic branches. The activation of new modes takes place in previously absolutely empty spectral space. Because of a considerable dispersion of the acoustic branches in comparison to the optical ones, and also due to the small damping of the phonons at small wave vectors, the SBMM folding peaks reveal very large $Q$-factors and are well distinguished in the spectra.

Figure 3.28 shows the emergence of the lines in a RbAg$_4$I$_5$ crystal during cooling. The temperature points at which an abrupt enrichment of the spectra with the new lines occurs, are well pronounced. Obviously, these are

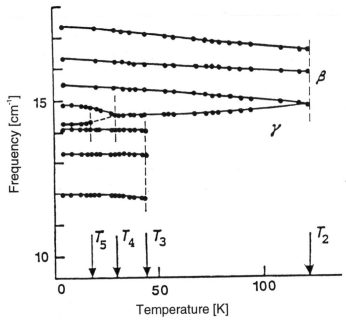

**Fig. 3.28.** Fine structure frequencies observed in $RbAg_4I_5$ $\nu s$ temperature. $T_2$ is a known $\beta$-$\gamma$ phase transition at 122 K. $T_3$, $T_4$, $T_5$ are unknown points

indications of crystal lattice structure changes and hence of the existence of the unknown structural phase transitions at these points. These BWO discoveries have been confirmed by other methods [3.42–44].

We would like to draw the reader's attention to the widths of the folding peaks. In some cases they do not exceed a few hundredths of a $cm^{-1}$. To register such lines the spectrometer should have a spectral resolution of up to $10^{-3}\,cm^{-1}$, which is entirely ensured by the BWO-spectrometers and highly conjectural for other techniques. We did not succeed, for example, in an attempt to extend the spectrum shown in Fig. 3.27b to higher frequencies using the commercial Fourier spectrometer Bruker IFS 113 V: the spectrum splitting was qualitatively observed but the lines were not completely resolved, thus there appeared to be many fewer. The same drawback is also characteristic of the Raman spectra [3.44].

It is very interesting that the acoustic nature of the folding peaks motivates a direct connection of the BWO investigations (which are infrared in their essence) to the ultrasonic ones. The universal and comparatively simple dispersion law of the acoustic branches with the sound velocity given by their slope allows one to reconstruct the dispersion of the branches in the whole Brillouin zone *via* the SBMM spectra [3.45, 46]. Hence, in this case the optical ($q = 0$) spectroscopy represented by the BWO measurements provides information generally accessible exclusively to neutron scattering technique.

### 3.2.6 Relaxors and Dipole Glasses

If one imagines the transition from a simple dielectric (rigid phonon spectrum) to a displacive ferroelectric (with a resonant soft mode) and then to an order-disorder ferroelectric (relaxation soft mode), the next step would be a so-called *relaxor*. Classical representatives of the relaxors are the systems of the type $(Pb,La)(Zr,Ti)O_3$ and $(Ba,Sr)Nb_2O_6$ [3.47], although the range of such objects is very broad. Normally these are mixed crystals (solid solutions) with a complicated composition manifesting strong temperature anomalies of the dielectric properties, widely spread out over the spectrum, and the temperature diffusion of the phase transitions. Strong flat losses and a smooth $\varepsilon'$-dispersion are characteristic of the SBMM range. Various relaxational forms with empirical stretching coefficients are typically used for their description.

On the phenomenological level the relaxor problem in dielectric spectroscopy is close to the problem of the *dipole glasses*. At low temperatures the dipole glasses come to a certain dynamic state giving a strong and wide, spectrally broadened dielectric response. Mixed crystal ADP-RDP is the model system in this field. Pure RDP and ADP crystals undergo ferroelectric ($T_C = 146\,K$) and antiferroelectric ($T_C = 148\,K$) phase transitions, respectively. In the RADP compound of middle concentrations, however, the phase transition is suppressed by the competition between the two differently directed types of polarizational ordering, and at low temperatures a dipole glass state without sharp anomalies evolves in place of one of the ordered phases.

Figure 3.29 shows our joined SBMM and infrared dielectric spectra of RADP [3.48]. They clearly reveal a distinguished soft mode behavior similar to that in pure RDP and ADP. But unlike the pure crystal, the evolution of the soft mode response with temperature in RADP is not interrupted by an abrupt phase transition. The transformation of the spectra continues monotonically down to the lowest temperatures.

Figure 3.30 shows the eigenfrequency $(2\pi\tau)^{-1}$ and the oscillator strength $S$ of the observed soft mode. The temperature behavior of both parameters is easily observed in the low-temperature phase beyond the paraphase. The most striking feature is a considerable decrease of $S$ at low temperatures, indicating a coupling of the soft mode to other low-frequency excitations. These excitations are the well-known microwave and radio-frequency loss bands widely accepted as representing the glass state dynamics in RADP [3.49, 50]. A remarkable property of these glass state excitations is again the lack of the oscillator strength conservation. Thus they appear to be the transmissive rollers in the global evolution of the dielectric response function. The probable scheme is sketched in Fig. 3.31. Phenomenologically the picture looks like the soft mode estafettical behavior finished by the coupling to the central peak. Well known in application to structural phase transitions, this phenomenon usually develops in a narrow temperature range in the vicinity of $T_C$. The RDP-ADP dielectric function behavior makes it reasonable to consider the

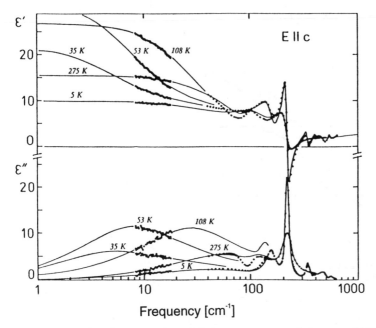

**Fig. 3.29.** Submillimeter-infrared dielectric spectra of the mixed RDP-ADP crystal. *Points* are the BWO and the infrared data, *lines* are the multioscillator fit

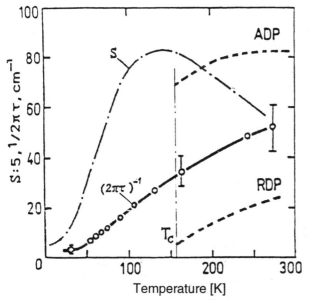

**Fig. 3.30.** Frequency $(2\pi\tau)^{-1}$ and oscillator strength $S$ *vs* temperature of the soft mode in RDP-ADP crystal. *Dashed* segments show $(2\pi\tau)^{-1}$ behavior in the pure RDP and ADP crystals

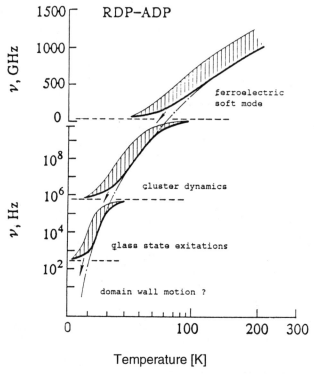

**Fig. 3.31.** Relay-race temperature behavior of the polar excitations in RDP-ADP crystal. *Thick solid line* shows excitations frequencies $\nu s$ temperature, while *shading* shows arbitrarily oscillator strengths

glassy state as some intermediate state during the phase transition process, stretching over many dozens of degrees, and still unfinished even at liquid helium temperature.

### 3.2.7 New Family of Ferroelectrics of TlGaSe$_2$ Type

We have discovered via BWO measurements ferroelectricity and the incommensurate phase in the family of a triple layer semiconductors of the TlGaSe$_2$ type [3.11].

Figure 3.32 shows the SBMM dielectric spectra of TlGaSe$_2$ obtained by the BWO-spectrometer. They reveal a typical ferroelectric soft mode with a linear temperature behavior of the squared frequency and with a divergence of its dielectric strength obeying the Curie–Weiss law. At $T = 120\,\mathrm{K}$ an essential change of the type of spectra is observed; it changes from the resonance (negative values of $\varepsilon'$) to relaxational behavior. At a slightly lower temperature of $T = 107\,\mathrm{K}$ an abrupt step-wise change of the whole picture in Fig. 3.32 is observed: the absorptivity and the dispersion of $\varepsilon'$ sharply decrease. The same

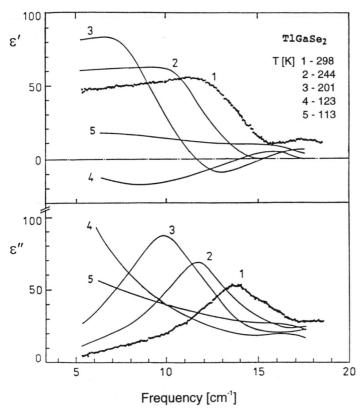

**Fig. 3.32.** Submillimeter dielectric spectra of TlGaSe$_2$ crystal at different temperatures. Temperature evolution of the spectra exhibits a typical ferroelectric soft mode behavior

properties with somewhat different characteristic temperatures are demonstrated by another crystal of this family TlInS$_2$. Complete analogy of that which is happening in TlGaSe$_2$ and TlInS$_2$ with the phenomena in known crystals, suggests the existence of incommensurate and ferroelectric phases for TlGaSe$_2$ and TlInS$_2$.

Much activity was aimed towards the verification of the SBMM data by various techniques. Anomalies of dielectric, optical, acoustical, and thermal properties, as well as of light and neutron scattering, were registered at the temperatures pointed out by us. A spontaneous polarization (ferroelectricity) was registered at low temperatures, and an incommensurate phase was found by neutron scattering [3.51, 52].

An interesting change of the TlGaSe$_2$ dynamics is observed when the selenium atoms are substituted for the sulfur atoms. These changes are shown in Fig. 3.33 *via* the behavior of the soft mode frequencies. The process of the softening is qualitatively similar for the crystals having different sulfur con-

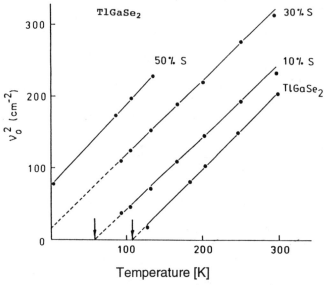

**Fig. 3.33.** Soft mode frequencies $\nu s$ temperature in the mixed $TlGa(Se, S)_2$ crystals. *Arrows* show the phase transition points

tent. However, there is an essential distinction shown by the different values of frequencies where the soft modes start their way towards zero frequency. As a result, only a few of them succeed in reaching zero frequency at a real (positive) temperature. As is seen, the phase transitions in the mixed crystals $TlGa(Se_xS_{1-x})_2$ are not realized at a sulfur content of more than 25 %. Obviously the frozen soft mode which did not reach zero frequency should reveal a giant dielectric response at low frequencies. The soft mode dynamics is very similar to the one described above in reference to the ADP-RDP crystal. By analogy, one may again assume that a dipole glass state is realized in the system $TlGa(Se_xS_{1-x})_2$ at low temperatures and at intermediate ratios of Se and S (60 and 30 %).

We believe that the family of layered crystals of the $TlGaSe_2$ type, which allow large-range atomic substitutions and have a large collection of interesting dynamic properties, may serve as promising model objects for study in the field of the physics of phase transitions.

### 3.2.8 Superionic Conductors

A strongly increased interest of researchers in the 1970s was detected towards crystalline dielectrics with high ionic conductivity – superionic conductors (solid electrolytes) [3.53]. In these amazing materials the atoms of one sort can relatively easily move in a rigid lattice formed by the atoms of another sort. The conductivity of the most outstanding superionics can reach values of $\approx 1\,\Omega^{-1}\,cm^{-1}$ at room temperature (10 orders of magnitude larger than in usual dielectrics) and is much higher than the electronic conductivity in these materials.

The superionics are interesting from various points of view and one of these concerns the peculiarities of the frequency-temperature behavior of their dielectric response function. The manifestation of all possible mechanisms of ionic transport could be expected in it: processes of diffusion, screening, temperature activation and localization of the charge carriers, and interaction of the movable ions with the crystal lattice and with each other. In a rough approximation, the experimentally observed frequency panorama of the response consists of two qualitatively different parts – the low-frequency part, where the conductivity has a purely diffuse character, and the high-frequency part (IR band) where the response becomes oscillatory. The transition between these two regimes falls in the SBMM range, where we have performed our investigations of the superionics by the BWO technique.

We have studied crystals of several different types (fourth column of Table 3.3). The SBMM data on the two most widely known compounds – AgI, and Na-$\beta''$–Al$_2$O$_3$ – are presented in Figs. 3.34, 35. Along with the details (which are interesting in each specific case) the spectra reveal the general feature of all superionics – a well-pronounced relaxation at the link between the diffusive and the oscillatory regimes. It manifests itself as a powerful anomalous dispersion ($\varepsilon'$ decrease $\nu s$ frequency) which appears in the SBMM

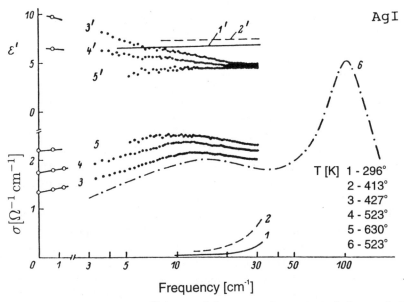

**Fig. 3.34.** Submillimeter $\varepsilon'(\omega)$ and $\sigma(\omega)$ spectra of superionic AgI crystal. *Points –* BWO data, *open circles –* MW data of H. Roemer et al., curve 6 – IR data of P. Bruesch et al.

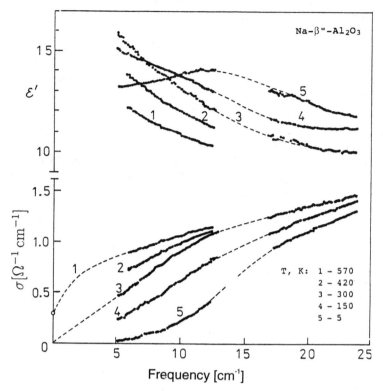

**Fig. 3.35.** Submillimeter $\varepsilon'(\omega)$ and $\sigma(\omega)$ spectra of superionic Na-$\beta''$-alumina crystal. *Dashed lines* are for the the eye. *Open circle* is the dc data

spectra and then disappears again during a linear increase of the temperature and a corresponding monotonous increase of the dc conductivity.

We have analyzed the situation in the framework of a simple dispersion model composed of the free carrier conductivity (the Drude formula), an oscillator (5), and a relaxator (6). The result regarding the number of the particles participating in the different types of motion appeared to be the most interesting. It unambiguously showed that the charges strongly interact with each other and that a significant role in the dynamics is played by the screening (backward currents) in the range of intermediate concentrations [3.54].

The discovery of the universal behavior of conductivity in superionics stimulated our search for a way to analytically describe the frequency dependent response, including its main features. For this purpose, following *Takagi* [3.55], we have generalized the Onsager equation to the case of conducting materials, and have obtained an expression for the conductivity which describes a continuous transition from the free movement of the carriers to their oscillatory motion [3.56]:

$$\sigma(\omega) = \frac{ne^2}{m}(1 - i\omega\tau_R)[\tau_R(\omega_0^2 - \omega^2) + \gamma - \gamma' - i\omega(1 + \gamma\tau_R)]^{-1},$$

where $n, e,$ and $m$ are the number, the charge, and the mass of the potentially movable particles, $\omega_0$ and $\gamma$ are the frequency and the damping of the oscillatory motion, and $\gamma' = \omega_0^2/\gamma$. The new parameter $\tau_R$, compared to the known models (5) and (6), is the residence time. It determines the time when the particle is localized. Formula (7) transforms continuously into the oscillator expression when $\tau_R \to \infty$, and into the Drude one when $\tau_R \to 0$. The results obtained from this model are described in [3.11, 56].

### 3.2.9 Electronic Conductors

The conducting materials – low dimensional conductors, conventional and high-$T_C$ superconductors, semiconductors (3, 7 and 8th columns of Table 3.3) – were studied by us in order to reveal the most general regularities of their response function behavior [3.11]. This topic of the spectroscopy of conductors and semiconductors seems to be the most weakly developed in the background of the enormous information devoted to the properties of the specific

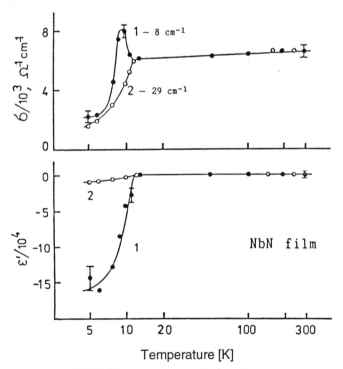

**Fig. 3.36.** Submillimeter spectra of $\varepsilon'$ and $\sigma \, vs$ temperature of superconducting NbN film. Lines are guides for the eye

materials studied in the specific conditions. This situation is due to a high degree of individuality of the investigated objects, and to the dependence of their properties on the growing technology. It is evident that from the fundamental viewpoint the samples of the most interest are only those which are comprehensively characterized by various methods.

The examples of the SBMM data on the conductivity obtained on the BWO-spectrometer are presented in Figs. 3.36, 37. Their main feature is that they show the absolute values of the real and imaginary parts of the dielectric response function. The SBMM spectra of conductors, as a rule, are rather impressive and very sensitive to the temperature. They allow, within definite models, for the determination of mobilities, concentrations, and effective masses of the charge carriers, and for specification of the scattering mechanisms.

In Fig. 3.36 the monotonic decrease of the conductivity of niobium nitride (NbN) at $29\,\mathrm{cm}^{-1}$ reveals reliably the opening of the energy gap in the excitation spectrum. It is interesting that the phenomenon looks more

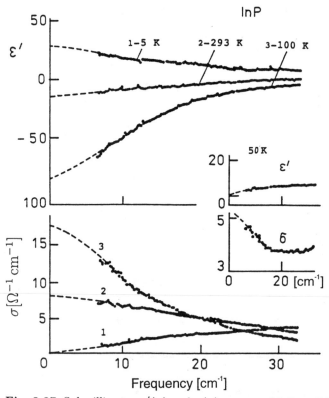

**Fig. 3.37.** Submillimeter $\varepsilon'(\omega)$ and $\sigma(\omega)$ spectra of InP at different temperatures. *Dashed lines* are the fit by additive Drude-relaxator model

complicated at lower frequencies: while going down below the critical temperature, the conductivity at $8\,\mathrm{cm}^{-1}$ first grows up and then decreases. This kind of temperature behavior of the conductivity is known as a *coherence peak*. Its occurrence in the NbN film suggests that the ground state in this material is of a singlet origin (3.57).

Figure 3.37 demonstrates the process of freezing out of the charge carriers in InP. A typical Drude-like dispersion mechanism observed at room temperature is replaced by a high-frequency Debye-like relaxation during cooling down. The nature of this phenomenon is not clear to us, at present.

### 3.2.10 Antiferromagnets

We have mastered our BWO technique for the magnetic measurements on the antiferromagnets yielding the SBMM values of the magnetic permeability $\mu'$ and $\mu''$ in addition to the $\varepsilon'$ and $\varepsilon''$ spectra. Fig. 3.38 shows the panorama of the ferro- and antiferromagnetic absorption in $YFeO_3$ recorded by the BWO-spectrometer. The narrow gaps on the background of the interference pattern for two orthogonal orientations are clearly observed. The resonance lines are extremely sensitive to the external magnetic field and orientation of the sample. We emphasize once more, here, the necessity of high intensity, resolution, and polarization of radiation. Let us stress, for example, that due to the high degree of polarization a very strong absorption line in the $h_c$ orientation at $18\,\mathrm{cm}^{-1}$ does not distort the oscillation pattern in the orthogonal $h_b$ orientation (see shaded domain).

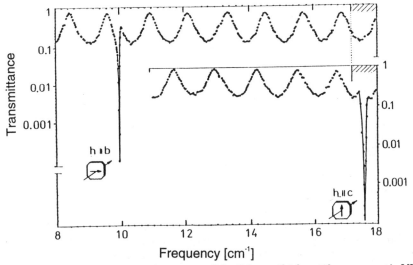

**Fig. 3.38.** Transmitance $vs$ frequency of the 0.8 mm thick antiferromagnetic $YFeO_3$ plate for two orthogonal orientations. Narrow resonances at 10 and $17.5\,\mathrm{cm}^{-1}$ are ferromagnetic and antiferromagnetic modes

In general, the job of $\mu'$ and $\mu''$ evaluation is highly complicated: when both $\varepsilon$, $\mu \neq 1$ then the complete set of the electrodynamic parameters $T$, $R$, $\phi$, and $\psi$ has to be measured. However, we have essentially simplified the problem by utilizing the fact that the magnetic resonances in the antiferromagnets are very sharp, and that the magnetic permeabilities are very close to unity except for the narrow part of the spectrum in the vicinity of the resonance line. In this particular case one can reliably separate the dielectric

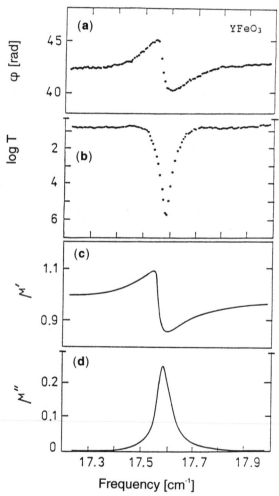

**Fig. 3.39 a–d.** Magnetic measurements on YFeO$_3$ crystal: (**a**) and (**b**) – phase $\phi(\omega)$ and transmission $T(\omega)$ spectra of 0.8 mm plate (BWO data); (**c**) and (**d**) – magnetic permeability $\mu'(\omega)$ and $\mu''(\omega)$ spectra, calculated from the above $\phi(\omega)$ and $T(\omega)$ data. Shown is the range of the antiferromagnetic resonance (*shaded* segment in Fig. 3.38)

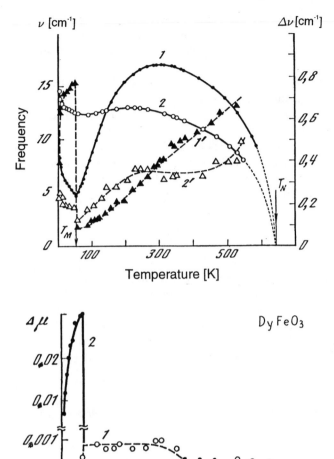

**Fig. 3.40 a,b.** Temperature dependence of parameters of AFMR modes in DyFeO₃: (a) – frequencies ($1 - \nu_1$, $2 - \nu_2$; and linewidths $1' - \Delta\nu_1$; $2' - \Delta\nu_2$); (b) – magnetic contributions ($1 - \Delta\mu_1$, and $2 - \Delta\mu_2$)

and magnetic measurements. We measured the spectra $\varepsilon'(\nu)$ and $\varepsilon''(\nu)$ of the orthoferrites outside the magnetic absorption lines by the usual method (part I of present review) and then interpolated this data in the resonance domains. The spectra of $\mu'$ and $\mu''$ were calculated with the known values of $\varepsilon'$ and $\varepsilon''$, either point by point directly from $T(\nu)$ and $T(\nu)$ spectra, or by fitting the measured spectrum $T(\nu)$ with a Lorentzian.

The high frequency fragment of the $T(\nu)$ added by the $\phi(\nu)$ spectrum, together with the $\mu'(\nu)$ and $\mu''(\nu)$ spectra evaluated from these two, is presented in Fig. 3.39. Figure 3.40 shows a typical example of the complex temperature evolution of the antiferromagnetic modes. The experimental material presented here was used for theoretical modeling of the magnetic phase transitions in orthoferrites [3.58].

## 3.3 Conclusion

We have described our experience with the application of BWOs for the characterization of devices and materials at SBMM waves and also for performing fundamental investigations in solid state physics. It is our opinion that for a long time such activity will continue to be timely and promising in obtaining many useful results.

At the same time, the following possible ways of broadening of this activity are clearly seen:

1) The BWO potentialities are pertinent not only for solid state physics but also for MW and IR spectroscopies, laser physics, plasma diagnostics, and astronomy.
2) The developed BWO techniques can be used, for the measurement of liquids. This direction is connected with applications in medicine and biology.
3) The ability of the BWO measurements to join the infrared and low-frequency radio measurements is promising in view of the development of the all-wavelength dielectric spectroscopy.
4) There is a possibility of considerable increase of the rapidity of the BWO measurements by using higher speed detectors, control elements, and new software. This opens the possibility of investigations of fast processes in substances, and of doing this under pulsed exposures.
5) Owing to the optical nature of the measurements, the comparatively large wavelength, and the high "quality" of the radiation, the BWO techniques are highly demonstrative, making them excellent educational equipment for beginning researchers (students) in the field of radiophysical measurements and coherent optics.

## Acknowledgements

During our activity we have collaborated with many people who have been our co-authors during different periods. The most significant contribution was made by our closest laboratory colleagues Yu.G. Goncharov, B.P. Gorshunov, S.P. Lebedev, Academician A.M. Prokhorov, and also Professor J. Petzelt in Prague. We are sincerely grateful to all of them. We also acknowledge the professional handling of the text of the manuscript by V.V. Voitsehovskii, the stimulation of the writing of this review and friendly support by Professor G. Gruner, and the interest taken the review and careful reading of the text by A. Schwartz.

We would like to thank The Russian Fundamental Research Foundation and The Soros Foundation for their financial support of these investigations during recent years.

# References

3.1    M.B. Golant, Z.T. Alekseenko, Z.S. Korotkova, L.A. Lunkina, A.A. Negirev, O.P. Petrova, T.B. Rebrova, V.S. Savel'ev: Pribory i Tekhnika Eksp. **3**, 231 (1965)

3.2    G. Convert, T. Yeou, P.C. Mouton: *Proc. 4th Int'l Congr. Microwave Tubes* (entrex Publishing, Eindhoven 1963) p. 739

3.3    E.M. Gershenson, M.B. Golant, A.A. Negirev, V.S. Savel'ev: Backward wave oscillators of millimeter and submillimeter wavelength ranges (Radio and Communication Press, Moscow 1985) [in Russian]

3.4    R.A. Valitov, S.F. Dubko, V.V. Kamishan, V.M. Kuzmichev, B.M. Makarenko, A.V. Sokolov, V.P. Sheiko: Technique for submillimeter waves (Soviet Radio Press, Moscow 1969) [in Russian]

3.5    D.H. Martin (ed.): *Spectroscopic Techniques for Far-Infrared. Submillimeter, and Millimeter Waves* (North-Holland, Amsterdam 1967)

3.6    N.A. Irisova: Vestnik AN SSSR **10**, 63 (1968)

3.7    E.M. Gershenson: Uspehi Fizicheskih Nauk **122**, 164 (1977)

3.8    V.N. Aleshichkin, V.V. Meriakri, G.A. Krahtmakher, E.E. Ushatkin: Pribory i Tekhnika Eksp. **4**, 150 (1971)

3.9    A.F. Krupnov: Vestnik AN SSSR **7**, 18 (1978)

3.10   A.A. Volkov, Yu.G. Goncharov, G.V. Kozlov, S.P. Lebedev, A.M. Prokhorov: Infrared Phys. **25**, 369 (1985)

3.11   G.V. Kozlov (ed.): Submillimeter dielectric spectroscopy of solids, Proc. Inst. General Phys., Vol.25 (Nauka, Moscow 1990) [in Russian]

3.12   G.V. Kozlov, A.M. Prokhorov, A.A. Volkov: Submillimeter dielectric spectroscopy of solids, in *Problems in Solid-State Physics*, ed. by A.M. Prokhorov (Mir, Moscow 1984)

3.13   M. Born, E. Wolf: *Principles of Optics* (Pergamon, Oxford 1970)

3.14   A.A. Volkov, G.V. Kozlov, S.P. Lebedev: Radiotechnika i Electronika **24**, 1405 (1979)

3.15   L.A. Vainshtein: Electronika Bol'shikh Mochnostey **2**, 26 (1963)

3.16   W.G. Chambers, C.L. Mok, T.J. Parker: J. Phys. **13**, 1433 (1980)

3.17   V.P. Shestopalov, L.M. Litvinenko, S.A. Masalov, V.G. Sologub: Diffraction of waves on grids (Kchar'kov, Moscow 1973) [in Russian]

3.18   U.R. Ulrich: Infrared Phys. **7**, 37 (1967)

3.19   A.A. Volkov, B.P. Gorshunov, A.A. Irisov, G.V. Kozlov, S.P. Lebedev: Int'l J. Infrared and Millimeter Waves **3**, 19 (1982)

3.20   R.N. Clarke, C.B. Rosenberg: J. Phys. E **15**, 9 (1982)

3.21   A.E. Kaplan: Radiotechnika i Electronika **10**, 1781 (1964)

3.22   B.P. Gorshunov, I.V. Fedorov, G.V. Kozlov, A.A. Volkov, A.D. Semenov: Solid State Commun. **87**, 17 (1993)

3.23   A.A. Volkov, Yu.G. Goncharov, B.P. Gorshunov, G.V. Kozlov, A.M. Prokhorov, A.S. Prokhorov, V.A. Kozhevnikov, A.M. Cheshitskii: Sov. Phys. - Solid State **30**, 988 (1988)

3.24   B.P. Gorshunov, A.A. Volkov, G.V. Kozlov, L. Degiorgi, A. Blank, T. Csiba, M. Dressel, Y. Kim, A. Schwartz, G. Grüner: Phys. Rev. Lett. **73**, 308 (1994)

3.25   C. Kittel: *Introduction to Solid State Physics* (Wiley, New York 1967)

3.26   H. Poulet, J.P. Mathieu: *Vibrational Spectra and Symmetry of Crystals* (Gordon & Breach, Paris 1970)

3.27   L.D. Landau, E.M. Lifshitz: *Electrodynamics of Condensed Media* (Pergamon, Oxford 1960)

3.28   A.A. Volkov, G.V. Kozlov, A.M. Prokhorov: Infrared Phys. **29**, 747 (1989)

3.29   A.D. Bruce, R.A. Cowly: *Structural Phase Transitions* (Taylor and Francis, London 1981)

3.30   R. Blinc, B. Zeks: *Soft Modes in Ferroelerctrics and Antiferroelectrics* (North-Holland, Amsterdam 1974)

3.31   M.E. Lines, A.M. Glass: *Principles and Applications of Ferroelectrics and Related Media* (Oxford Univ. Press, Clarendon 1977)

3.32   A.S. Barker Jr.: Phys. Rev. B **12**, 4071 (1975)

3.33   J. Petzelt, G.V. Kozlov, A.A. Volkov: Ferroelectrics **73**, 101 (1987)

3.34   G.V. Kozlov, E.B. Kryukova, S.P. Lebedev, A.A. Sobyanin: Ferroelectrics **80**, 233 (1988)

3.35   G.V. Kozlov, E.B. Kryukova, S.P. Lebedev, A.A. Sobyanin: Sov. Phys. - JETP **67**, 1689 (1988)

3.36   F. Jona, G. Shirane: *Ferroelectric Crystals* (Pergamon, Oxford 1962)

3.37   H.Z. Cummins, A.P. Levanuk (eds.): *Light Scattering near Phase Transitions* (North-Holland, Amsterdam 1983)

3.38   A.S. Barker Jr., J.J. Hopfield: Phys. Rev. A **135**, 1732 (1964)

3.39   A. Janner, T. Janssen: Europhys. News **13**, 1 (January 1982)

3.40   J. Petzelt. Phase Transitions **2**, 155 (1981)

3.41   A. Horioka, A. Sawada, R. Abe: Ferroelectrics **36**, 347 (1981); ibid **66**, 303 (1986)

3.42   A. Shawabkeh, J.F. Scott: J. Raman Spectrosc. **20**, 277 (1980)

3.43   I.H. Akopian, D.N. Gromov, B.V. Novikov: Fizika Tverdogo Tela **29**, 1475 (1987)

3.44   B.H. Bairamov, N.V. Lichkova, V.V. Timofeev, V.V. Toporov: Fizika Tverdogo Tela **28**, 1543 (1987)

3.45   A.A. Volkov, Yu.G. Goncharov, G.V. Kozlov, V.I. Torgashov, J. Petzelt, V. Dvorak: Ferroelectrics **109**, 363 (1990)

3.46   S. Kamba, V. Dvorak, J. Petzelt, Yu.G. Goncharov, A.A. Volkov, G.V. Kozlov: J. Phys. C **56**, 4401 (1993)

3.47   G. Burns, F.H. Ducol: Ferroelectrics **104**, 25 (1990)

3.48   A.A. Volkov, G.V. Kozlov, S.P. Lebedev, A.V. Sinitskii, J. Petzelt: Sov Phys. - JETP **74**, 133 (1992)

3.49   E. Courtens: Phys. Rev. B **52**, 69 (1984)

3.50   H.J. Brukner, E. Courtens, H.-Unruh: Z. Physik B **73**, 337 (1988)

3.51   R.A. Aliev, K.R. Allakhverdiev, A.I. Baranov, N.R. Ivanov, R.M. Sardarly: Fizika Tverdogo Tela **26**, 1271 (1984)

3.52   S.B. Vakhrushev, V.V. Zhdanova, B.E. Kvjatkovskii, N.M. Okuneva, K.R. Allakhverdiev, R.A. Aliev: Sov. Phys. - JETP Lett. **39**, 291 (1984)

3.53   M.B. Salamon (ed.): *Physics of Superionic Conductors*, Topics Curr. Phys., Vol. 15 (Springer, Berlin, Heidelberg 1979)

3.54   A.A. Volkov, G.V. Kozlov, J. Petzelt, A.S. Rakitin: Ferroelectrics **81**, 211 (1988)

3.55   Y. Takagi: J. Phys. Soc. Jpn. **47**, 567 (1979)

3.56   A.A. Volkov, G.V. Kozlov, A.S. Rakitin: Sov. Phys. - Solid State **36**, 189 (February 1990)

3.57   M. Tinkham: *Introduction to Superconductivity* (McGraw-Hill, New York 1975)

3.58   A.M. Balbashov, G.V. Kozlov, A.A. Mukhin, A.S. Prokhorov: Submillimeter spectroscopy of antiferromagnetic dielectrics, in *High Frequency Processes in Magnetic Materials*, ed. by G. Srinivasan, A. Slavin (World Scientific, Singapore 1995) Chap. 2, pp. 56-98

# 4. Waveguide Configuration Optical Spectroscopy

George Grüner
With 32 Figures

## List of Symbols

| | |
|---|---|
| $\sigma$ | Conductivity |
| $\sigma_1$ | Real part of the conductivity |
| $\sigma_2$ | Imaginary part of the conductivity |
| $q, \omega$ | Wavevector and frequency |
| $E$ | Electric field |
| $j$ | Electric current |
| $\epsilon$ | Dielectric constant |
| $\epsilon_1$ | Real part of the dielectric constant |
| $\epsilon_2$ | Imaginary part of the dielectric constant |
| $\delta_0$ | Skin depth |
| $\delta_{c_1}$ | Skin depth of cavity walls |
| $c$ | Speed of light |
| $D$ | Displacement vector |
| $N = n + \mathrm{i}k$ | Refractive index |
| $Z_\mathrm{s}$ | Surface impedance |
| $R_\mathrm{s}$ | Surface resistance |
| $X_\mathrm{s}$ | Surface reactance |
| $Z_0$ | Impedance of free space |
| $R$ | Reflectivity |
| $A, \phi$ | Amplitude and phase of the electromagnetic wave |
| $S$ | Scattering parameter |
| $Q$ | Quality factor of cavity |
| $W$ | Stored electromagnetic energy |
| $S'$ | Absorbed electromagnetic energy |
| $H$ | Magnetic field |
| $B$ | Magnetic induction |
| $\mu_0$ | Bohr magneton |
| $\lambda$ | Wavelength of electromagnetic radiation |
| $\Phi$ | Complex phase $= \phi + i \log(A/20)$ |
| $Y$ | Admittance |
| $Q$ | Quality factor |
| $f_0$ | Resonant frequency (unperturbed) |
| $f_\mathrm{s}$ | Resonant frequency (sample inserted) |

Topics in Applied Physics, Vol. 74
**Millimeter and Submillimeter Wave Spectroscopy of Solids** Ed.: G. Grüner
© Springer-Verlag Berlin Heidelberg 1998

| $\Gamma$ | Resonance width |
|---|---|
| $\omega_0^*$ | Complex resonance frequency of the cavity |
| $U$ | Electromagnetic energy of cavity |
| $g, \xi$ | Resonator constants |
| J | Bessel fuction |
| $d$ | Diameter of circular cavity |
| $h$ | Height of cavity |
| $s'$ | Sensitivity |
| $F$ | Source frequency |
| $A(f_0)$ | Transmitted power |
| $V$ | Network voltage |
| $I$ | Network current |
| $R_{1,2}$ | Reflection coefficients |
| $T_{1,2}$ | Transmission coefficients |
| $X, Y$ | Arguments of Bessel functions |
| $\phi_\mathrm{m}$ | Metallic shift |
| $\Delta f'$ | Shift of resonant frequency relative to shift for infinite conductivity |
| $N_s$ | Depolarization factor |
| $\sigma_a, \sigma_b$ | Conductivities in different directions |
| $\gamma_v$ | Specific heat |
| $m^*$ | Effective mass |
| $m_\mathrm{b}$ | Bandmass |
| $N$ | Number of electrons |
| $\mathbf{k}_\mathrm{F}$ | Fermi wavevector |
| $\chi$ | Susceptibility |
| $\tau$ | Relaxation time |
| $\tau^*$ | Enhanced relaxation time |
| $\omega_\mathrm{p}$ | Plasma frequency |
| $I_{1,2}$ | Conductivity sum rule integral |
| $\Delta$ | Superconducting, or density wave gap |
| $E, K$ | Elliptic integrals |
| $k_\mathrm{B}$ | Bolzmann constant |
| $\lambda_0$ | Penetration depth |
| $p_1$ | Amplitude of charge density wave |
| $\phi'$ | Phase of charge density wave |
| $S_1$ | Amplitude of spin density wave |
| $\lambda'$ | Electron–phonon coupling constant |
| $v_F$ | Fermi velocity |
| $q$ | Phason wavevector |
| $\mathcal{L}$ | Lagrangian |
| $\kappa$ | Elastic constant of density waves |
| $K$ | Restoring force |

The micro and millimeter wave spectral range (from approximately 1 GHz to 200 GHz (or, from approximately $3 \times 10^{-2}$ to $7\,\mathrm{cm}^{-1}$) represents the electromagnetic spectrum between radio frequency and the far infrared. This range is (or can be) defined by the experimental techniques. While rf components are used at lower frequencies and standard optical configurations at higher frequencies, in the frequency range defined above, the various components of the conductivity or dielectric constant are usually measured using waveguide components in either nonresonant or resonant configurations. The techniques utilized in this spectral range have under gone significant development in recent years. Due to improved sources and detectors, experiments up to the very far infrared range are feasible even under cryogenic conditions.

The fundamental purpose of the experiments is to determine the various components (real and imaginary parts) of the insertion loss; either a bridge or a resonant cavity can be utilized. From the various components of the loss, together with an appropriate knowledge of the sample geometry, the real and imaginary components of the complex conductivity

$$\sigma(\omega) = \sigma_1(\omega) + \mathrm{i}\sigma_2(\omega) , \qquad (4.1)$$

or alternatively the complex dielectric constant can be evaluated. In general, the advantage of the method over conventional optical experiments lies mainly in the fact that both components of the conductivity are directly measured. The main disadvantages of the measurements are the high cost associated with the apparatus and the fact that separate experiments, often employing different components (e.g., cavities), must be conducted at different frequencies.

Measuring the complex electrical conductivity as a function of the frequency is not a trivial task, and different problems are encountered in different frequency ranges. At very low frequencies ($< 1\,\mathrm{GHz}$) one probes the conductivity by attaching leads to the sample, and the specimen is represented by an RC circuit.

At higher frequencies several problems are encountered when such a measurement configuration is employed. Stray capacitances and standing waves become important, and also the dimensions of the specimens may become comparable to the wavelength – at least in the upper end of the millimeter wave spectral range. Consequently, the propagation of the electromagnetic fields is achieved by using waveguide configurations, and the specimen is incorporated in the structure, either in a resonant or in a nonresonant configuration. The frequency-dependent conductivity is then evaluated from the resulting perturbations of the electromagnetic fields.

This chapter focuses on the experimental techniques available in the millimeter wave spectral range, and on the analysis required to extract the components of the complex conductivity. Experiments on materials with known conductivity will be reviewed first; this serves the purpose of demonstratsing the feasibility and effectiveness of the experimental techniques, both for good and moderate conductors, as well as insulators.

The spectral range discussed plays an ever increasing role in condensed matter physics, as the frequency scale corresponds to energies comparable to energies associated with single particle and collective excitations of various solids. As examples of these experiments of highly correlated metals will be reviewed in their so-called Fermi liquid and in their various broken symmetry ground states, which includes the low energy electrodynamics of the super-conducting state and the dynamics of states called the charge density wave and spin density wave.

## 4.1 Optical Constants and Measured Parameters

The approach usually taken in describing the salient features of optical spectroscopy is to describe the electromagnetic fields in terms of the Maxwell equations and to account for the medium in terms of frequency-dependent complex optical conductivity or frequency-dependent complex dielectric constant.

Assume that we have a time and spatially dependent electric field of the form of

$$E(r, \omega) = E_0 \exp(iqr - \omega t) \,. \tag{4.2}$$

We also assume that in response to this field an electric current is induced, and this is given by $j(r - \omega)$. The parameter we want to establish is the so-called optical conductivity

$$\sigma(\omega) = \frac{j(r, \omega)}{E(r, \omega)} \,, \tag{4.3}$$

where $E$ and $j$ are the local fields and electric currents, with the conductivity given as

$$\sigma(\omega) = \sigma_1(\omega) + i\sigma_2(\omega) \,, \tag{4.4}$$

where $\sigma_1$ is the real part and $\sigma_2$ is the imaginary part. In the vast majority of cases there is a local relation between the current electric field as given by (4.3) and the conductivity has only a frequency but no spatial dependence. Cases where a nonlocal relation applies and the wavevector dependence is important (this is the case for the anomalous skin effect and for superconductors with a long coherence length) will not be discussed. We can also define a complex permittivity, or dielectric constant

$$\varepsilon(\omega) = \varepsilon_1(\omega) + i\varepsilon_2(\omega) = \varepsilon_\infty + 4\pi \frac{i\sigma(\omega)}{\omega} \,, \tag{4.5}$$

where $\varepsilon_\infty$ the so-called high-frequency dielectric constant.

Obviously, in the majority of cases where the conductivity is high, the response to a time-dependent field is a current which is spatially non-uniform with screening currents located near to the surface of the specimens; however the local relation between the current and electric field still applies. Our goal is to take the spatial dependence of the electric field and current into account and evaluate the conductivity as given above, in terms of relation (4.3). This is done by defining a set of optical constants and, consequently, a set of experimentally accessible parameters.

Assume that we apply a slowly varying time-dependent field to a specimen with contacts applied at the ends. The purpose of the contacts is to allow the electric charge to flow in and out of the specimen and thus to prevent the buildup of charges at the specimen boundary. The wavelength $q$ is, for small frequencies (say, below microwave) significantly larger than the specimen dimensions, and also, the conductivity is such that the electric field is not screened and is uniform throughout the specimen. The wavevector dependence of the problem can be neglected under such circumstances, but, by definition, is perpendicular to the electric field as required by the general principles of electrodynamics. There is a time lag between the electric field and current; this is expressed by the complex conductivity.

With increasing frequencies two things happen. The alternating electric current may be screened by the induced currents and progressively flows in a surface layer. This surface layer is determined typically by the parameter called the skin depth, which, at low frequencies (in the so-called Hagen-Rubens limit, where $\sigma_1 \ll \sigma_2$) is given by

$$\delta_0 = \frac{c}{(2\pi\varepsilon_1\omega)^{1/2}},\tag{4.6}$$

where $c$ is the speed of light.

For typical metals at room temperature the skin depth at a frequency of $10^{10}$ Hz is of the order of $10^{-3}$ cm, and is larger for many of the materials which will be discussed later. If this length scale is smaller than the dimension of the specimen, the effect has to be taken into account, and then the concept of so-called surface impedance becomes useful.

If the dimensions of the sample and the skin depth $\delta_0$ is also larger than the dimensions of the specimen, then the electric field inside is given by

$$\boldsymbol{E} = \frac{\boldsymbol{E}_0}{1 + N(\varepsilon - 1)},\tag{4.7}$$

where $N$ is the depolarization factor. One finds a homogeneous electric field only for certain sample geometry (such as a ellipsoid). The electric field as given above leads to an electric current – which will also be uniform only in case $\boldsymbol{E}$ – also does not have a spatial variation throughout the specimen. The absorbed energy

$$S' = \int_V \boldsymbol{j} \cdot \boldsymbol{E}dV,\tag{4.8}$$

where the integration is over the sample dimensions.

When the dimensions of the specimen are larger than the skin depth, the electric fields and currents are not uniform inside the specimen. Let us define a parameter called the surface impedance,

$$Z_{\mathrm{s}} = \frac{E_{z=0}}{\int\limits_0^\infty j(z)dz} = R_{\mathrm{s}} + iX_{\mathrm{s}}, \qquad (4.9a)$$

where $z = 0$ defines the surface, with the specimen occupying the space for $z > 0$, $j(z)$ is the spatially dependent electric current which decays exponentially with increasing $z$. Because of the local relation between the current and electric field, the spatial dependence of $j(z)$ is the same as the spatial dependence of $E$. The electric field $E_{z=0}$ refers to the magnitude of the transmitted wave $E_t$ just inside the specimen. The real Part $R_{\mathrm{s}}$ is the surface resistance and the imaginary part $X_{\mathrm{S}}$ the surface reactance. The tangential magnetic field at the surface

$$H_t = \frac{4\pi}{c} \int\limits_0^\infty j(z)dz. \qquad (4.9b)$$

At low frequencies we neglect the displacement current and therefore

$$\nabla \times \boldsymbol{H} = \frac{4\pi}{c} \boldsymbol{j}. \qquad (4.10)$$

Then (4.9) becomes

$$Z_s = \frac{4\pi E_t}{c \int_S (\nabla \times \boldsymbol{H})dS} = \frac{4\pi E_t}{c} \Big/ \int_t \boldsymbol{H} dl = \frac{4\pi}{c} \frac{E_t}{H_t}. \qquad (4.11)$$

Here, both $E_t$ and $H_t$ refer to fields at the surface, and we have used Stokes' theorem to convert a surface integral to a line integral which is parallel to $\boldsymbol{H_t}$ at the surface and extends into the metal well beyond the skin depth, where the magnetic field is zero. We can also use the relation

$$\nabla \times \boldsymbol{E} + \frac{1}{c} \frac{\partial \boldsymbol{B}}{\partial t} = 0$$

to obtain

$$\frac{\partial E_t}{\partial z} = \frac{i\omega}{c} H_t, \qquad (4.12)$$

which leads to another expression of the surface impedance:

$$Z_s = \frac{4\pi i\omega}{c^2} \left( \frac{E_t}{\partial E_t / \partial z} \right). \qquad (4.13)$$

Using the equation for the propagation of light in the medium we have

$$\frac{\partial E_t}{\partial z} = -\frac{i\omega}{c} E_t (n + ik) = \frac{i\omega}{c} E_t N \qquad (4.14)$$

giving

$$Z_s = R_s + iX_s = \frac{4\partial i\omega}{c^2} \frac{c}{i\omega N} = -\frac{1}{c} \left( \frac{4\pi i\omega}{\sigma} \right)^{1/2}. \qquad (4.15a)$$

Note: if rationalized MKS units are used instead of the Gaussian units, the surface impedance is given by

$$Z_s = \left( \frac{-i\mu_0\omega}{4\pi\sigma} \right)^{1/2}, \tag{4.15b}$$

this impedance being normalized to the impedance of free space $Z_0 = 377\,\Omega$. In accordance with convention, MKS units will be used in subsequent chapters, where the experiments are discussed and analyzed.

The energy loss is given by the real part of the Poynting vector:

$$S = \frac{c}{4\pi}|\boldsymbol{E} \times \boldsymbol{H}| = \frac{c}{4\pi}E_t H_t = \frac{E_t^2}{Z_s}. \tag{4.16}$$

The two parameters $R_s$ and $X_s$ are related to the changes of the scattering parameters and changes of the resonance characteristics when a specimen is placed in a waveguide, or forms part of the resonant cavity, these relations will be discussed in subsequent chapters.

The reflected power (the quantity which is usually observed by optical experiments, see for example, *Wooten* [4.1] is

$$R = \left( \frac{N-1}{N-1} \right)^2 = \frac{(n-1)^2 + k^2}{(n+1)^2 + k^2}. \tag{4.17}$$

Using (4.15) the reflected power is given, in terms of the surface impedance as

$$R = \left( \frac{Z_0 - Z_s}{Z_0 + Z_s} \right)^2, \tag{4.18}$$

where $Z_0$ is the impedance of free space. The reflectivity also can be written as

$$R = 1 - \frac{4R_s Z_0}{Z_0^2 + |Z_s|^2 + 2R_s Z_0}. \tag{4.19}$$

For highly conducting specimens $R_s$ and $Z_s \ll Z_0$, the reflectivity is determined by the surface resistance,

$$R \simeq 1 - \frac{4R_s}{Z_0}. \tag{4.20}$$

As will be discussed later, two approaches have been taken in order to extract the components of the optical conductivity. Either one measures the amplitude $A$ and phase $\phi$ of the reflected or transmitted power where the specimen is placed in a waveguide or the frequency shift and change of the quality factor $Q$ of a resonant structure containing the specimen. The set of two parameters gives $R_s$ and $X_s$, and then using (4.15) the components of the conductivity can be evaluated. Alternatively, one measures only $R_s$ [roughly proportional to changes in $A$ or in $(1/Q)$] and then this parameter can be combined, by using (4.19) to obtain the reflectivity $R$, which is combined with optical experiments conducted at higher frequencies $R(\omega)$ over a broad spectral range. The Kramers–Kronig relations can then be used to obtain $\sigma_1(\omega)$ and $\sigma_2(\omega)$.

## 4.2 Experimental Arrangements

A variety of experimental arrangements have been employed to extract the components of the complex conductivity, and we classify them as resonant and nonresonant methods. Both methods have obvious advantages and disadvantages, and both have been used effectively to study various phenomena. In this section we discuss the experimental arrangements which enable us to measure the complex impedance $Z_S$ associated with the specimen. The impedance is related to the components of the optical conductivity (and to the geometry of the specimen). This relation is, in general, fairly involved and depends on whether the sample is well or poorly conducting, and whether the real or imaginary part of the conductivity is larger. These questions will be discussed in Sect. 4.3.

### 4.2.1 Nonresonant Methods

The name refers to techniques where a nonresonant configuration is employed with the specimen inserted in the configuration. In general, one measures the change of the amplitude and the phase of the reflected or transmitted wave, and using these two parameters, the components of the complex conductivity can be extracted. The advantage of the method is that experiments can be conducted at various frequencies; the disadvantage lies in the fact that – at least for highly conducting specimens – the sensitivity is not high.

a) **Post in a Waveguide.** In general, one measures the change of the amplitude $A$ and phase $\phi$ of the transmitted or reflected wave in a waveguide with a specimen placed, as a post, at a well-defined position. These changes are referred to as $A_s$ and $\phi_s$. One then defines a scattering parameter

$$S = -10^{A_s/20} e^{i\phi_s} \, , \tag{4.21}$$

and in terms of the (complex) $S$, the impedance $Z$ is associated with the specimen given by

$$\frac{Z}{Z_0} = \frac{1+S}{1-S} \, , \tag{4.22}$$

where $Z_0$ is the impedance of the waveguide.

One also requires a well-defined electric or magnetic field configuration, and this can be achieved, among others, in a way shown in Fig. 4.1a. Here the waveguide is terminated by a short (conducting wall). Then the electric field has a maximum at the positions $x = \frac{3}{4}\lambda n$ where $\lambda$ is the wavelength of the

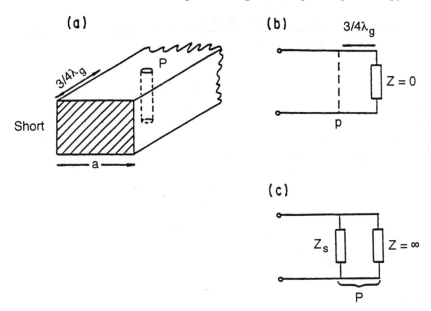

Fig. 4.1 a–c. Description of waveguide analysis. (a) Sample location. (b) and (c) are equivalent circuits for a sample $3\lambda/4$ from the shorting plate and with the shorting plate transformed to the sample position, respectively

radiation in the waveguide, $n = 1, 2 \ldots$, and the specimen is placed at this position. The equivalent circuit for this configuration is shown in Fig. 4.1b; and the circuit for the shorting plate transformed to the sample position is displayed in Fig. 4.1c. With this transformation the reflected amplitude and phase is related to $Z_s$.

Usually, changes in $A$ and (in particular for highly conducting specimens) $\phi$ are small. To circumvent this disadvantage, bridge configurations can be employed.

b) **Bridge Configurations.** A bridge is an interferometric device, the layout is shown in Fig. 4.2. Power from the source is split into two components, one traversing the reference arm, and the other the sample arm, and recombined to provide a detected output, determined by the relative conditions of the two arms. The wave traveling the reference arm is reflected from a fixed short. In the sample arm, the wave is reflected from the unknown impedance which is to be measured. The bridge is nulled (zero detected output) by adjusting an attenuator and a phase meter. At this condition, the attenuations in the two arms are equal and the phase shifts differ by an integral multiple of $2\pi$.

The measurement technique consists of nulling the bridge with and without the sample and noting the changes $A_s$ and $\phi_s$, in attenuation and phase readings. Without the sample, the terminating impedance is a short. The impedance due to very thin needle-like samples is a pure shunt, which we

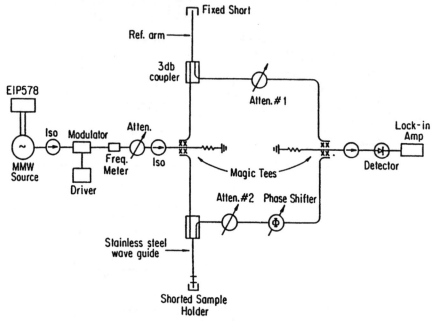

**Fig. 4.2.** Diagram of the microwave bridge, discussed in the text

call $Z_{in}$. With the sample, the terminating impedance is $Z_{in}$, referred to the plane at the sample position.

The specimen leads to change in the amplitude and phase of the transmitted signal which is interpreted as follows: If we define the change in the complex phase as $\delta\Phi = \phi_s + i\log(A_s/20)$, then the ratio of the scattering parameter sample in, $S_{in}$, to the scattering parameter sample out, $S_{out}$, is

$$S_{in}/S_{out} = \exp(\delta\Phi).\qquad(4.23)$$

In general, the scattering parameter is given by $S = (1-Y)/(1+Y)$, where $Y$ is the admittance. It follows that

$$\tan\left(\frac{-\Delta\Phi}{2}\right) = \frac{(Y_{in}Y_{out})/Y_0}{1 + Y_{in}Y_{out}/Y_0^2},\qquad(4.24)$$

where $Y_{in}$ and $Y_{out}$ are the admittances with and without the sample (sample in and sample out), respectively. The form above is invariant to transformations down the transmission line and may be evaluated at any position. For simplicity we chose to evaluate the admittances at the sample position. The short must be transformed by a distance $3 - \lambda/4$ (defining the positive direction towards the shorting plate) and becomes an open circuit ($Y_{out} = 0$). Determining $Y_{in}$ is, in general, a complicated calculation, but for a sample approximated as a circular cylinder of radius $R$ traversing the waveguide with the long axis parallel to the electric field, the admittance can be calculated. This will be discussed in Sect. 4.3.

## 4.2.2 Resonant Methods

**a) Resonant Cavities: General Formalism.** Resonant cavities offer a high sensitivity detection configuration for the measurement of complex impedance. For a high quality factor $Q$, the electromagnetic field scatters on the specimen (which is part of, or contained in the structure) roughly $Q$ times, enhancing the sensitivity enormously. Typical $Q$ values, for cavities constructed in the millimeter wave range are displayed in Fig. 4.3.

The most practical means of developing a high frequency, high $Q$ resonator is to enclose the fields within a body whose dimensions are comparable to the desired wavelength of operation. Such a device is normally referred to as a resonant cavity, and it will support a series of modes, with a lower cut-off, each corresponding to a unique distribution of fields. Most resonators are fabricated out of highly conducting material, and in this case, the fields are determined solely by the boundary conditions at the interior surfaces. For ease of fabrication, one is generally limited to cylindrical or rectangular cavities, and in these cases one can easily determine the field distribution with the resonator. For a given measurement frequency, the minimum cavity dimensions are roughly given by $1/2$ the wavelength in each physical dimension.

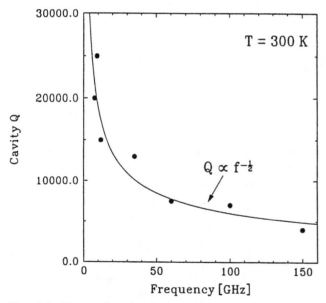

**Fig. 4.3.** Measured quality factor of circular enclosed cavities at various frequencies utilized in the authors laboratory. The *full line* is a curve calculated as described in the text

The $Q$ of a cavity is determined by the energy loss per cycle and is defined as

$$Q \equiv \frac{f_0}{\Gamma} = \frac{\omega_0 \langle W \rangle}{L} ,\qquad (4.25)$$

where $f_0$ is the resonant frequency, $\Gamma$ the full width of the resonance of half-maximum, $\langle W \rangle$ is the time-averaged energy stored in the cavity

$$\langle W \rangle = \frac{1}{16\pi} \int_{V_c} (|\boldsymbol{E}(\boldsymbol{r})|^2 + |\boldsymbol{H}(\boldsymbol{r})|^2)dV ,\qquad (4.26)$$

and $L$ the energy loss per cycle. There are three main loss mechanisms: Ohmic losses in the cavity walls, radiation losses in the coupling device, and losses within the sample placed in the cavity. These different mechanisms contribute in an additive way and the total loss is determined by the suin of the reciprocals of the $Q$ related to each loss mechanism,

$$\frac{1}{Q} = \frac{1}{Q_u} + \frac{1}{Q_r} + \frac{1}{Q_s} ,\qquad (4.27)$$

where $Q$ represents the total, or loaded, cavity loss; $Q_u$ the Ohmic, or unloaded, loss; $Q_r$ the radiative loss, and $Q_s$ the sample loss.

Apart from a constant of the order of unity (determined by the particular mode), $Q_u$ is given by the ratio of the volume occupied by the field to the volume of the conductor into which the fields penetrate (*Jackson* [4.2]).

$$Q_u = \frac{V_c}{A\delta} ,\qquad (4.28)$$

for a nonmagnetic ($\mu = \mu_0$) conductor where $A$ is the total interior surface area and $\delta$ the skin depth. Maximizing the cavity volume to surface area ratio generally increases the $Q$.

The relation between the cavity parameters and the surface impedance can also be established in a more rigorous fashion (see for example *Klein* et al. [4.3]).

The relation between the complex resonant frequency and the electric and magnetic fields in the cavity [4.4–6] is given by

$$\frac{\omega^* - \omega_0^*}{\omega^*} = -\frac{\int_{\Delta V}[(\varepsilon - \varepsilon_0)\boldsymbol{E}_0^*\boldsymbol{E} + (\mu - \mu_0)\boldsymbol{H}_0^*\boldsymbol{H}]dV}{\int_V (\varepsilon_0 + -\boldsymbol{E}_0^*\boldsymbol{E}_0 + \mu_0\boldsymbol{H}_0^*\boldsymbol{H}_0)dV} .\qquad (4.29)$$

In this expression $\omega_0^*$ is the complex resonant frequency for the unperturbed cavity of volume $V$ whose permittivity and permeability are $\varepsilon_0$ and $\mu_0$, and $E_0$ and $H_0$ are the unperturbed electric and magnetic fields. After a sample characterized by $\varepsilon$, $\mu$, and volume $\Delta V$ ($\varepsilon$ and $\mu$ can be complex quantities) is introduced into the cavity, the fields $E$, $H$ and the frequency are perturbed and are designated by quantities without the subscripts.

First, consider a cavity of volume $V$ enclosed by an infinitely conducting surface. When excited by an oscillating electromagnetic wave, the cavity will resonate at a frequency $\omega_0$ determined by the boundary conditions for the fields. For a perfectly conducting surface, the tangential component of the electric field and the normal component of the magnetic induction must be zero (i.e., $n \times E = 0$ and $n \cdot B = 0$ where $n$ is the unit normal vector). With these constraints imposed, only certain eigenfrequencies are allowed for a given set of cavity dimensions as discussed by *Montgomery* [4.7], *Jackson* [4.2], or *Ramo* et al. [4.8]. Also, since there are no losses in this system, the quality factor $Q$ is infinite. For a cavity with a finite surface impedance, the fields penetrate the walls as the effective shielding by the inducted currents is reduced. The resulting change in energy, which is stored primary in the magnetic field, alters the resonant frequency, and the Ohmic losses in the surface produce a finite $Q$. The effect can be described quantitatively again using (4.29).

Assuming that the fields outside the sample are the same as the unperturbed fields, (4.29) becomes

$$\frac{\omega^* - \omega_0^*}{\omega^*} - \frac{i}{2Q_u} = -\frac{\int_{\Delta V} \left( (\varepsilon - \varepsilon_0) E_0^* E + (\mu - \mu_0) H_0^* H \right) dV}{4U_{total}} \qquad (4.30)$$

[see (4.26)]. Here the subscripted symbols (e.g., $\varepsilon_0$, $E_0$, $H_0$, and $\omega_0$) are the unperturbed quantities pertaining to an ideal cavity whose boundaries have zero surface impedance, the quantities lacking subscripts refer to the perturbed cavity resulting from a nonzero $Z_s$. The integral is over the additional volume acquired when the fields are allowed to penetrate into the walls because of the finite surface impedance. This volume is approximately equal to the area of the cavity times the skin depth. At places where the electric field is parallel to the surface, the unperturbed boundary field is zero, while on surfaces where it is perpendicular, the free charges completely screen out the interior electric fields. In either case there is no electric field contribution to the integral in (4.30).

Recalling that the unperturbed $Q$ is infinite, the complex resonant frequencies on the left-hand side are

$$\omega^* = \omega - i\frac{\omega}{2Q_u} \quad \text{and} \quad \omega_0^* = \omega_0 . \qquad (4.31)$$

Substituting expression (4.31) into the left-hand side of (4.30) and dropping the second-order terms yields

$$\frac{\omega - \omega_0}{\omega} - \frac{i}{2Q_u} = \frac{1}{4U_{total}} \int_{\Delta V} (\mu - \mu_0) H_0^* \cdot H \, dv . \qquad (4.32)$$

Now $(\mu - \mu_0) H$ is the difference between the magnetic induction below and above the surface of the cavity. Hence the integrand approximately represents the magnetic energy density within the conducting walls, and it is proportional to the increase in magnetic energy due to a finite surface impedance. In (4.32) we take $\mu_0 \to 0$, and as for the unperturbed case, it is assumed that no energy is stored in the walls. Since the material is nonmagnetic, we can then set $\mu = \mu_0$, and (4.32) then reduces to

$$\frac{\omega - \omega_0}{\omega} - i\frac{1}{2Q_u} = -\frac{\mu_0}{4U_{\text{total}}} \int_{\Delta V} \boldsymbol{H}_0^* \cdot \boldsymbol{H} dv , \tag{4.33}$$

which can be written as

$$\frac{\omega - \omega_0}{\omega} - i\frac{1}{2Q_u} = -\frac{\mu_0}{4U_{\text{total}}} \int_s \boldsymbol{H}_0^* \left( \int_0^\infty H_z dz \right) ds . \tag{4.34}$$

In (4.33) the volume integral is expressed as integrals perpendicular to and over the surface area of the cavity.

Using (4.12), one obtains (note that $c = 1/\sqrt{\varepsilon_0 \mu_0}$ )

$$E(0) = -\mu_0 \omega \int_0^\infty H_t dz , \tag{4.35}$$

and combining this equation with (4.13) the surface impedance becomes

$$Z_s = -\frac{i\mu_0\omega}{H_t(z=0)} \int_0^\infty H_t dz . \tag{4.36}$$

Incorporating (4.36) into (4.34) one obtains

$$\frac{\omega - \omega_0}{\omega} - i\frac{1}{2Q_u} = i\frac{Z_s}{4\omega U_{\text{total}}} \int_s |H_0|^2 ds . \tag{4.37}$$

Since

$$U_{\text{total}} = \frac{1}{2} \int_V |H_0|^2 dv , \tag{4.38}$$

(4.37) can also be written as

$$\frac{\omega - \omega_0}{\omega} - i\frac{1}{2Q_u} = -i\frac{Z_s}{2\omega} \frac{\int_s |H_0|^2 ds}{\int_V |H_0|^2 dv} . \tag{4.39}$$

We can separate this equation into its real and imaginary parts. Expressed more conveniently,

$$\Delta\omega = \omega - \omega_0 = \sum_j \frac{1}{2}\gamma_j X_{sj} , \tag{4.40}$$

$$\Gamma = \frac{\omega_0}{Q_u} = \sum_j \gamma_j R_{sj} , \tag{4.41}$$

where the terms on the right are summed over all surfaces enclosing the resonant cavity, and $R_{sj}$ and $X_{sj}$ are, respectively, the resistance and reactance of the $j^{\text{th}}$ surface. The coefficient $\gamma_j$ referred to as the resonator constant, is defined by

$$\gamma_j \equiv \frac{1}{\mu_0} \frac{\int_s |H_0|^2 ds}{\int_V |H_0|^2 dv} . \tag{4.42}$$

With expression (4.42), one is able to relate measurable parameters such as $\Delta\omega$ and $\Gamma$ to sample dependent quantities like $R_s$ and $X_s$.

**b) Cavity Perturbation: The Principle.** In a cavity perturbation experiment one measures the change in width $\Gamma$, $\Delta\Gamma$, and frequency $\omega_0$, $\delta\omega_0$ of a resonant structure which occurs due to the introduction of a foreign body into the resonant structure. If the perturbation is suitably small, one can determine the material properties of the body from the measured changes in the cavity characteristics. One configuration commonly employed is a cavity endplate measurement, where one wall of the resonant cavity is replaced by the sample under investigation. The change of the resonant frequency and quality factor is measured, and from these two parameters the impedance associated with the specimen, and subsequently, the conductivity, can be evaluated. For the other method, one places the specimen inside the resonant structure. The cavity perturbation equation is

$$\Delta\hat\omega = -4\pi\gamma\hat\alpha\,, \tag{4.43}$$

where $\hat\omega = \omega_0 - i\Gamma/2Q$, $\gamma$ is a constant which is proportional to the sample to cavity volume ratio $V_s/V$ – times a constant which depends on the field configuration within the cavity, and $\hat\alpha$ is the sample polarizability.

**Resonant Cavity Measurement Configurations.** Depending on the sample size and shape, different types of perturbations are made, see for example, *Donovan* et al. [4.9] and references cited therein. In some cases a particular type of perturbation is chosen to maximize the measurement sensitivity, while in other cases the physical characteristics of the sample to be investigated dictate the arrangement. An isotropic sample of the appropriate size will give identical results for each of the following perturbation techniques. However, as the direction of the induced currents differ with each technique, anisotropic samples will generally give different results in each configuration.

If the sample dimensions exceed the cavity diameter in at least two directions, the endplate can be replaced by the sample, as illustrated in Fig. 4.4. In this, the so-called endplate technique, certain other restrictions are placed on the properties of the sample.

If the sample dimension is much smaller than the wavelength, one can introduce the sample inside the cavity. The maximum sensitivity is realized at the anti-node of either the electric or magnetic field. The antinode positions, $H_{max}$, and $E_{max}$ are shown in Figs. 4.5, 6 for a cavity in the $TE_{011}$ mode. In a different configuration (Fig. 4.7) the sample, resting on a platform, can be rotated in and out of the cavity during measurement.

With both the endplate and enclosed perturbation techniques, the cavity must be opened to introduce the sample. This is usually accomplished by removing the endplate. The sample is then positioned atop a small quartz rod which is glued to the endplate at the appropriate location of either $E_{max}$ or $H_{max}$ (Figs. 4.5, 6). The quartz rod does not significantly alter the cavity $Q$ but does introduce a slight shift in the resonance frequency.

The major disadvantage of the enclosed perturbation is the unavoidable presence of a frequency offset which accompanies the opening of the cavity.

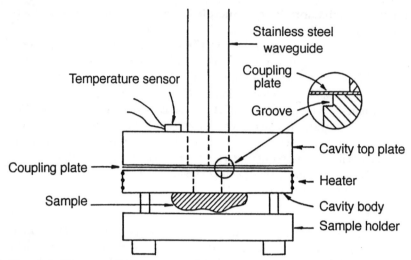

**Fig. 4.4.** Diagram of the setup used for the end plate measurement of the surface impedance. The end plate has been replaced by the sample under investigation

One way to avoid this problem would be to mechanically introduce the sample through a small hole in the cavity. While such a technique would remove the unknown frequency offset, the additional hole in the cavity would serve to increase the radiation losses $(1/Q_r)$. However, it is expected that at low frequencies, where the volume to surface area ratio of the cavity is large, such a hole would not drastically reduce the $Q$.

**c) Resonant Cavity Measurement Techniques.** The experimental methods used in a cavity perturbation measurement vary somewhat, but in all cases, two separate measurements, one with the sample and one without it (in the case of the endplate technique, a copper end late replaces the sample), are performed at each temperature and, except for the in situ perturbation, two complete temperature sweeps must be made. Between the temperature sweeps, the cavity is opened and the sample is removed. The empty cavity is closed and the temperature cycle is repeated.

In order to extract the cavity characteristics $\Gamma$ and $f_0$, two different techniques are generally employed.

**Width Technique.** The most standard method employed to measure $f_0$ and $\Gamma$ uses a broad sweep of the source frequency over a range which is at least several times the resonance width. The detected signal is amplified and then averaged (typically 100 times) with a digital oscilloscope before fitting with a Lorentz curve. A schematic diagram of the setup is shown in Fig. 4.8.

There are several factors which complicate the measurement of the frequency. The resonant frequency $f_0$ of the cavity is strongly temperature-dependent (due to the thermal contraction of the cavity), and this requires one to either periodically adjust the source central frequency or to make a

## TE$_{011}$ cylinder cavity
### H fields

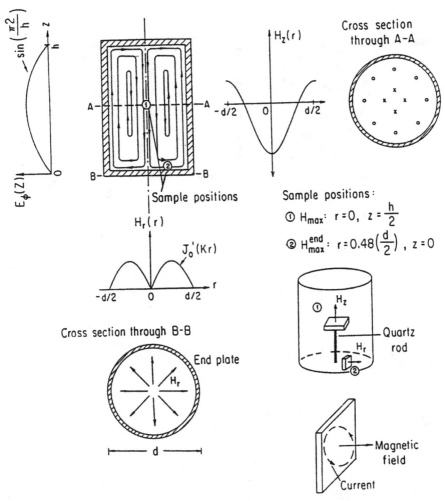

**Fig. 4.5.** Magnetic field lines inside a cylindrical cavity resonating in the TE$_{011}$ mode. Both the maximum magnetic field in the cavity $H_{max}$, position 1, and on the end plate $H_{max}^{end}$, position 2, are indicated on the figure

sweep which is larger than the shift caused by the thermal contraction of the cavity. Typically, a change of 500 MHz to 2 GHz is seen in $f_0$ between 1.2 K and 300 K, and such a large frequency sweep, together with the smearing of the peak due to the averaging, makes it difficult to attain a high precision measurement.

# TE₀₁₁ cylinder cavity
## E fields

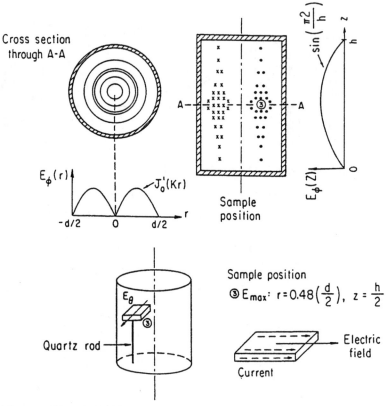

**Fig. 4.6.** Electric field lines inside a cylindrical cavity resonating in the TE₀₁₁ mode. The position of the maximum electric field $E_{max}$, position 3, is also indicated on the figure

**Amplitude Technique.** A more accurate way to measure both the bandwidth and the characteristic frequency in a cavity in the transmission mode, uses a narrow frequency sweep about $f_0$ while measuring the power transmitted at the resonance. As the sweep in frequency is typically orders of magnitude smaller than with the conventional method, the characteristic frequency can be directly measured with a frequency counter up to an arbitrary accuracy by appropriately averaging. The bandwidth is inversely related to the square of the power absorbed at the resonance (the integral of the transmitted spectrum is proportional to the power of the source times $1/\Gamma$) and can thus be determined indirectly. Using this type of detection necessitates that the source central frequency is locked to the characteristic frequency of the resonator. This can be achieved with a feedback loop, in a fashion similar

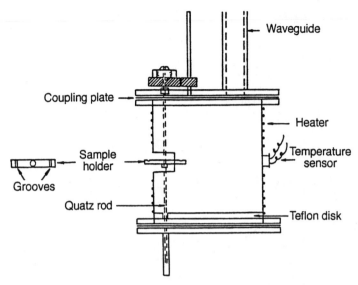

**Fig. 4.7.** Cutaway view of a cylindrical cavity with the rotating sample holder. The sample is placed at either position 1 or position 2 on the teflon sample tray. In this configuration, the sample will pass through $E_{max}$

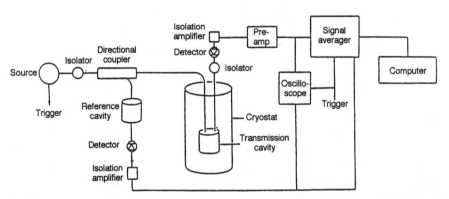

**Fig. 4.8.** Conventional detection scheme for experiments using the width method. The source is swept over a broad frequency range and the detected signal is then fit to a Lorentzian curve by a computer

to that used in a standard FM radio. We refer to this measurement technique as the amplitude technique, and the principle of the technique is displayed in Fig. 4.9.

In order to lock on to the resonance, one must constantly minimize the difference between the central frequency of the source, $F$, and $f_0$. This can be achieved by periodically modulating, with period $T$, the source frequency by a small amount $\delta F$ about $F$, while making a phase sensitive detection

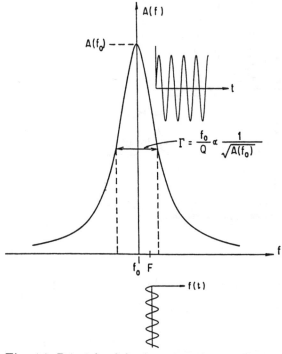

**Fig. 4.9.** Principle of the detection scheme called "amplitude method"; a modulation frequency $f(t)$ is superimposed to the central frequency $F$ of the source. The inphase component of the detected signal is proportional to the difference $(F - f_0)$. By accurately measuring $F$ (with a frequency counter) and the amplitude of the signal at $f_0$, $A(f_0)$, we can deduce both $f_0$ and $\Gamma_0$

of the in-phase component of the transmitted signal at the frequency $1/T$. The in-phase component measures the derivative of the absorption spectrum and, to first order, this is proportional to the error signal $(F - f_0)$. This error signal is fed back to the source and constantly provides a correction to the source frequency. In this way the source frequency is locked to the cavity resonance frequency. With the above mentioned sources, $F$ is proportional to the total dc voltage $V_{dc}$ applied, and changes of $F$ in steps less than 1 ppm are possible. The total frequency of the source $F_s$ is the sum of the central frequency $F$ and the time-dependent modulation $f(t)$

$$F_s = F + f(t). \tag{4.44}$$

The modulation, $f(t)$ is periodic (typically 10 ms) with an amplitude (approximately 0.1 MHz) which is less than 1 bandwidth. A computer aided program monitors the error voltage and adjusts $F$ to minimize the feedback voltage, thus providing the large long term frequency shifts required to keep up with a strongly temperature-dependent resonant frequency. This system is equivalent to the Automatic Frequency Control (AFC) in an EPR spectrometer.

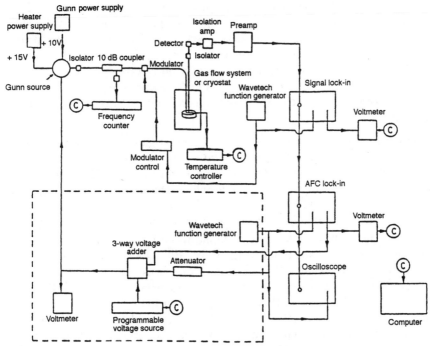

**Fig. 4.10.** Experimental setup for the amplitude measurement technique. This configuration is used for a varactor-tuned Gunn oscillator. If an Impatt source is used, then the dashed box is replaced by an HP 8530B mainframe

The short term integrator is set by the phase-sensitive detector and is of the order of a few seconds. The long term integration is done by an interfaced computer, that adjusts periodically (few minutes period) the voltage $V_{dc}$ to minimize the error signal.

To measure the bandwidth, the emitted microwave power is chopped and the transmitted power is directly measured with an additional phase sensitive detector. A schematic of the setup can be seen in Fig. 4.10.

### 4.2.3 Evaluation of the Complex Conductivity from the Measured Impedance

In this section we review the different theoretical aspects which are needed in order to extract the intrinsic material properties of a sample under investigation. Several authors have previously studied this problem of cavity perturbation in a variety of limiting cases: *Champlin* and *Krongard* [4.10], together with *Brodwin* and *Parsons* [4.11], solved the problem exactly for a sphere placed in the maximum of either the magnetic or electric field; *Buranov* and *Shchegolev* [4.12] examined the case of a prolate spheroid in the electric field maximum under the condition that the electromagnetic radiation uniformly within the sample (depolarization regime); *Cohen* et al. [4.13] together with work by *Ong* [4.14] investigated a prolate spheroid in the electric field maximum in which the electromagnetic radiation was confined to a small volume near the surface of the sample (skin depth regime).

The dielectric and conducting post placed in a waveguide was first considered by *Schwinger* and *Saxon* [4.15] and the solutions are used extensively to extract the components of the complex conductivity or complex dielectric constant.

**Discontinuities in Waveguides.** Several investigators [4.15–17] calculated the effects a discontinuity has on the field configurations inside a waveguide; here we follow the method used by *Schwinger* and *Saxon* [4.15]. We will begin by examining a rectangular waveguide where the discontinuities (e.g., the sample) are cylindrical about the $y$-axis. By calculating the complex power flowing down the guide using Polyning's Energy theorem and equating it to $\frac{1}{2}V_nI_n^*$, the normalization factor $C$ is found to be $\left(\frac{2}{a}\right)^{1/2}$. Consequently, it is possible to represent each mode in the waveguide as a transmission line with a characteristic impedance $Z_n$ and propagation constant $\kappa_n$.

The post in a waveguide we will consider is shown in Fig. 4.11a. One can use an equivalent circuit [4.18, 19] to represent the disturbances a discontinuity produces in the dominant mode. Since the field equations are linear, the discontinuity can be represented by a circuit equation for a four terminal network:

$$V_1 = Z_{11}I_1 + Z_{12}I_2, \tag{4.45a}$$

$$V_2 = Z_{21}I_1 + Z_{22}I_2, \tag{4.45b}$$

where $V_1$, $V_2$, $I_1$, and $I_2$ are defined in Fig. 4.11b. The impedances $Z_1$, $Z_2$, $Z_1$, and $Z_2$ are determined by the structure of the obstacle and do not depend on the terminal condition. The impedance matrix satisfies two important requirements: first, the reciprocity condition holds (i.e., $Z_{12} = Z_{21}$), and second, a dissipationless system can be described by a purely reactive network (i.e., $Z_{k1} = -IX_{k1}$). Hence all statements pertaining to the field (i.e., the distributed circuit elements) have an analogous expression in terms of conventional electrical quantities (i.e., lumped circuit elements), and the full power of network theory can be applied. If we also assume that the obstacle is symmetric with respect to the $z$-axis, $Z_{11} = Z_{22}$. Taking symmetry and reciprocity conditions into account, the final expressions relating the voltages to the currents are

$$V_1 = Z_{11}I_1 + Z_{12}I_2, \tag{4.46a}$$

$$V_2 = Z_{12}I_1 + Z_{11}I_2, \tag{4.46b}$$

and these equations represent the conventional $T$ network shown in Fig. 4.11c.

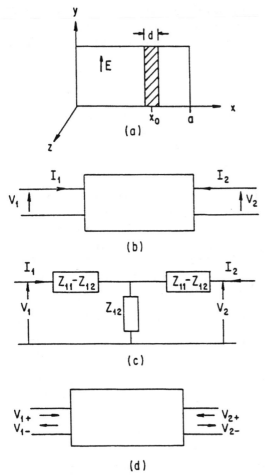

**Fig. 4.11.** (a) Diagram of a rectangular waveguide with the cross-hatched area representing the (inductive) discontinuity. (b) Network representation of the discontinuity. (c) Conventional $T$ network. The various impedances are calculated in the text. (d) The reflected and incident voltages at parts 1 and 2 in the scattering matrix formulation

An alternative to the impedance formulation of a transmission line is the scattering matrix representation. This is important because measured quantities are usually associated with this representation. In this case, one concentrates on the traveling incident and reflected voltage waves at each of the ports of the network as defined in Fig. 4.11d. The components of the scattering matrix $S$ are the given by

$$V_{1-} = R_1 V_{1+} + T_1 V_{2+} , \tag{4.47a}$$

$$V_{2-} = T_2 V_{1+} + R_2 V_{2+} , \tag{4.47b}$$

where $R$ and $T$ are the reflection and transmission coefficients for each of the two ports (i.e., $\boldsymbol{V}_- = \overleftrightarrow{S})\boldsymbol{V}_+$. Symmetry considerations allow us to write $R_1 = R_2$ and $T_1 = T_2 = T$. The two representations are not independent, however, and can be related using

$$S = \frac{Z - Z_0}{Z + Z_0} \quad \text{or} \quad \frac{Z}{Z_0} = \frac{1+S}{1-S}. \tag{4.48}$$

$Z_0$ is the characteristic impedance of the waveguide. By measuring the amplitude and phase of the incident and/or the reflected or transmitted waves, the scattering coefficients can be computed. Equation (4.48) is then used to relate the scattering coefficients to the impedance.

Ultimately, we wish to know the value of the complex conductivity of a material. In order to accomplish this, the complex conductivity must be related to the effective impedance of the sample in the waveguide for a given specific geometry. The configuration most often chosen is to place a thin, rod-shaped sample across the narrow dimension of the rectangular waveguide (i.e., an inductive geometry) as displayed in Fig. 4.11a. *Schwinger* and *Saxon* [4.15] used a variation technique to find the impedance for this configuration. For the even and odd cases are the following:

$$Z_{11} + Z_{12} = -i\frac{a}{\lambda_g}\csc^2\left(\frac{\pi x_0}{a}\right)\left[\ln\left[\frac{4a}{\pi d}\sin\left(\frac{\pi x_0}{a}\right)\right] - 2\sin^2\left(\frac{\pi x_0}{a}\right) - \frac{1}{4}\left(\frac{\pi d}{\lambda}\right)^2\right]$$

$$+ 2\sum_{n=2}^{\infty}\sin^2\left(\frac{n\pi x_0}{a}\right)\left\{\left[n^2 - \left(\frac{2a}{\lambda}\right)^2\right]^{-1/2} - \frac{1}{n}\right\}$$

$$+ \frac{J_0(Y)}{J_0(X)}\left(\frac{1}{XJ_0(Y)J_1(X) - YJ_0(X)J_1(Y)}\right), \tag{4.49a}$$

$$Z_{11} - Z_{12} = -i\frac{a}{\lambda_g}\sin^2\left(\frac{\pi x_0}{a}\right)$$

$$\times \frac{\left(\frac{\pi d}{a}\right)^2}{\frac{1}{2}X^2\left(\frac{J_1(Y)}{J_1(X)}\right)\left(\frac{1}{XJ_0(X)J_1(Y) - YJ_0(Y)J_1(X)}\right) - 1}, \tag{4.49b}$$

where

$$X = \frac{\omega d}{\lambda}, \quad \text{and} \quad \sqrt{\varepsilon}X = \left(1 + i\frac{\sigma}{\varepsilon_0\omega}\right)^{1/2}\left(\frac{\pi d}{\lambda}\right)$$

and $J_n$ is the Bessel function of order $n$, [4.17]. $\lambda_g$ is the wavelength in the waveguide, $\lambda$ is the free space wavelength, and $a$ and $d$ are defined in Fig. 4.11a. In the above equations, $\sigma$ can be complex. If the sample is at the center of the waveguide, then $x_0 = a/2$. Also, in most cases, the sample diameter, $d$, is small compared to the wavelength, and $\sqrt{\varepsilon}kd \ll 2$ to neglect the impedances in the arms of the $T$ circuit (i.e., $Z_{11} \approx Z_{12}$; the effective impedance reduces to a pure shunt configuration with a value given by

$$Z = Z_\infty + Z_\sigma \,, \tag{4.50a}$$

$$
\begin{aligned}
Z_\infty &= -i\frac{a}{2\lambda_g}\left[\ln\left(\frac{4a}{\pi d}\right) - \frac{1}{4}\left(\frac{\pi d}{\lambda}\right)^2\right.\\
&\quad \left. + 2\sum_{n=3}^{\infty}\left\{\left[n^2 - \left(\frac{2a}{\lambda}\right)^2\right] - 1/2 - \frac{1}{n}\right\}\right],
\end{aligned} \tag{4.50b}
$$

$$Z_\sigma = -i\frac{a}{2\lambda_g}\frac{J_0(Y)}{J_0(X)}\left(\frac{1}{XJ_0(Y)J_1(X) - YJ_0(X)J_1(Y)}\right). \tag{4.50c}$$

In this expression the summation is over the odd terms. The first term is a purely reactive contribution due to the sample configuration, and the second term, which may be complex, depends on the sample's conductivity. As seen from (4.50b), $Z_\infty$ is inductive. However, for samples that do not transverse the waveguide or do not make electrical contact with the wall, a capacitive configurational contribution in series with the inductive term may be present. The capacitance is probably a result of charge build up at the ends of the sample, and the relative importance of these two contributions is determined by the sample size.

Eventually, we wish to evaluate the complex conductivity of the sample from the measured impedance. Unfortunately, (4.50) is cumbersome, and can only be inverted numerically. It is possible, however, to examine several limiting cases that lead to explicit expressions for the conductivity.

First consider the case of a metallic sample in the skin depth regime where

$$\sigma_i = 0 \quad \text{and} \quad \delta_{cl} = \left(\frac{2}{\mu_0\omega\sigma_r}\right)^{1/2} < d.$$

The argument of the Bessel function containing the dielectric constant becomes

$$Y \approx \pi d\lambda\left(-i\frac{\sigma_1}{\varepsilon_0\omega}\right)^{1/2} = \frac{\pi d}{\lambda}\left(\frac{\sigma_1}{2\varepsilon_0\omega}\right)^{1/2}(i+1). \tag{4.51}$$

This quantity is much greater than one, allowing us to use the asymptotic form for the Bessel function which is

$$J_n(Y) = \left(\frac{2}{\pi Y}\right)^{1/2}\cos\left(Y - \frac{n\pi}{2} - \frac{\pi}{4}\right). \tag{4.52}$$

Letting

$$\frac{\pi d}{\lambda}\left(\frac{\sigma_1}{2\varepsilon_0\omega}\right)^{1/2} = A,$$

the zeroth order reduces to

$$J_0(Y) \sim \left(e^{i(A-\pi/4)}e^{-A} + e^{-i(A-\pi/4)}e^A\right) \approx e^{-i(A-\pi/4)}e^A, \tag{4.53}$$

since the second term dominates when $A \gg 1$. A similar expression is found for $J_1(Y)$. The opposite limit must be assumed for the Bessel functions with argument $X = \pi d/\lambda$. When $X \ll 1$, only the leading term of the small argument expansion needs to be retained, and it is given by

$$J_n(X) = \frac{1}{\Gamma(n+1)} \left(\frac{X}{2}\right)^n ,$$ (4.54)

where $\Gamma(n+1)$ is the gamma function. In this limit,

$$J_0(X) = 1 \quad \text{and} \quad J_1(X) = \frac{X}{2} .$$ (4.55)

Substituting these limiting expressions into (4.50) yields

$$Z_\sigma = \frac{a}{2\lambda_g} \frac{\delta_{c1}}{d} (1 - i) .$$ (4.56)

The equivalent circuit associated with $Z_\sigma$ is an inductance and resistance in series.

Next, let's examine the situation where $\sigma$ is complex. If the complex conductivity is not too large, such that

$$|\varepsilon| = \left[\varepsilon_r^2 + \left(\frac{\sigma_1}{\varepsilon_0 \omega}\right)^2\right]^{1/2} \ll \left(\frac{\lambda}{\pi d}\right)^2$$

(this is satisfied provided $\varepsilon_r \ll (\pi d/\lambda)^2$ and $d \ll \sqrt{2}\delta_{c1}$) then all the Bessel functions may be replaced with the leading term of their small argument expansions given by (4.54). For this case, (4.50c) reduces to

$$Z_\sigma = \frac{\alpha}{\sigma_1 + i\sigma_2} ,$$ (4.57)

where

$$\alpha = \left(\frac{a}{\lambda_g}\right) \left(\frac{\lambda}{\pi d}\right)^2 \varepsilon_0 \omega .$$ (4.58)

The equivalent circuit consists of a resistance, $R_s = \alpha/\sigma_1$, in parallel with a reactance, $X_s = -\alpha\sigma_i$, and this combination is in series with $Z_\infty$. Also, the sign of the dielectric constant determines whether the sample's reactive contribution is inductive or capacitive.

When using the complex impedance bridge, the phase $\phi$ and attenuation $A_s$ are the actual quantities measured. Hence, their dependence on the sample's intrinsic parameter is important. To further explore this point, we will consider the bridge response for a lossy wire whose diameter is less than the skin depth and $\sigma_i = 0$. The effective impedance has a resistive component given by (4.57) in series with $Z_\infty$, and it is related to the scattering coefficient, $S = 10^{A_s/20} e^{i\phi}$, by (4.48). Defining $Z_\infty = iX_\infty$ and $R_\infty = \alpha\sigma_1$, these expressions can be combined to yield

$$A_s = -10 \log_{10} \left(P^2 + Q^2\right) ,$$ (4.59a)

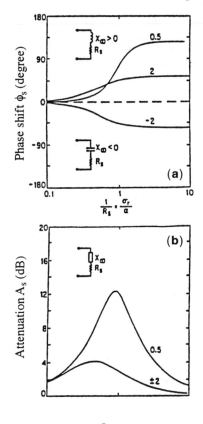

**Fig. 4.12.** A graph of the (a) phase shift $\phi_s$ and (b) measured attenuation $A_s$, as a function of $1/R_s = \alpha_1/\alpha$ for $X_\infty = 0.5$ and $\pm 2$. It is assumed that $\sigma_1 = 0$ [4.20]

$$\tan(\phi_s) = \frac{Q}{P}, \qquad (4.59b)$$

where

$$P = \frac{R_s^2 + X_\infty^2 - 1}{(R_s + 1)^2 + X_\infty^2} \quad \text{and} \quad Q = \frac{2X_\infty}{(R_s + 1)^2 + X_\infty^2}.$$

A graph of $A_s$ and $\phi_s$ versus $1/R_s$ for $X_\infty = \pm 2$ and 0.5 can be found in Fig. 4.12. For very low conductivities, both the loss and phase shift are small. As the conductivity increases, $A_s$ initially rises linearly with $\sigma_1$. Eventually, $A_s$ starts to drop, producing a peak which occurs at approximately $1/R_s = 1$ of $|X_\infty| \ll 1$ or at $1/R_s \approx 1/X_\infty$ in the opposite limit. Approaching the metallic side, $A_s$ decreases at a rate proportional to $1/\sigma_1$ before entering the skin depth regime where $A_s \propto 1/\sqrt{\sigma_1}$. The magnitude of the phase shift, which is small for insulating samples, increases where the attenuation peaks and saturates at a value determined solely by $X_\infty$. This metallic shift is given by

$$\tan\left(\frac{1}{2}\phi_m\right) = \frac{1}{X_\infty}. \qquad (4.60)$$

### 4.2.4 Cavity Perturbation

In this method one measures the cavity characteristics separately both before (0) and after (s) a sample has been inserted into a cavity, or replacing part of the cavity walls. The change in the complex frequency is

$$\Delta\hat{\omega} = \hat{\omega}_s - \hat{\omega}_0 . \tag{4.61}$$

(In the following, subscript s refers to a configuration with the specimen, 0 without the specimen). If the change $\Delta\hat{\omega}$ is adiabatic, then the product of the period and the time averaged energy stored is invariant [4.21]

$$\frac{\langle W\rangle}{\hat{\omega}} = \text{constant} . \tag{4.62}$$

This implies that

$$\frac{\Delta\langle W\rangle}{\langle W\rangle} = \frac{\Delta\hat{\omega}}{\hat{\omega}} \sim \frac{f_s - f_0}{f_0} - \frac{i}{2}\left(\frac{1}{Q_s} - \frac{1}{Q_0}\right) , \tag{4.63a}$$

$$\langle W\rangle = \frac{1}{16\pi} \int_{V_e} \left(|\boldsymbol{E}(\boldsymbol{r})|^2 + |\boldsymbol{H}(\boldsymbol{r})|^2\right) , \tag{4.63b}$$

$$\Delta\langle W\rangle = -\frac{1}{4} \int_{V_s} (\boldsymbol{P}\boldsymbol{E}^* + \boldsymbol{M}\boldsymbol{H}^*)dv . \tag{4.63c}$$

In our definition, $\Delta$ is the variation caused by the introduction of a foreign body in the resonating structure; $\Delta f$ is the frequency shift, and $\Delta\Gamma$ the change in the width of the resonance.

**a) Cavity End-Plate Replacement.** The experiment is performed by clamping a sample, which has a flat polished surface, to the bottom of the cavity and measuring the resonant parameters ($f_0$ and $Q_L$) as a function of temperature. After replacing the sample with a smooth copper end plate, the sample out resonance is recorded at the same temperatures. By adding up the contributions of each loss mechanism existing in the cavity, the total nergy dissipated is found and expressed in terms of a loaded quality factor which can be determined from the resonance width.

In principle, the presence of the sample only changes the unloaded $Q_u$, and the change in width caused by the sample is

$$\Delta\Gamma = \Delta(1/Q)\gamma_e(R_s - R_{cu}) , \tag{4.64a}$$

$$\Delta\omega = \frac{1}{2}\gamma_e(X_s - X_{cu}) . \tag{4.64b}$$

Here the sample and copper end plates are assumed to have the same surface area with the resonator constants $\gamma_e$ [see (4.40, 41)].

For the end plate of a cavity in the $TE_{011}$ mode, the resonator constant was calculated and one obtains [4.3]

$$\gamma_e = \frac{2}{\mu_0 h^3} \left( \frac{c_0 \pi}{\omega} \right)^2 . \tag{4.65}$$

Because the sample and copper end plates may not be tightened down to the same level, the difference in the geometrical contributions given by (4.63 b) usually does not vanish, and this is accounted for with a finite difference frequency $\Omega$. Even though there may be an uncertainty in the absolute value of the reactance, its temperature dependence can be found if $\Omega$ is taken to be constant. Also observe that in (4.64) both the resistance and reactance are measured relative to that of copper. Care must be exercised in calculating the copper contributions, especially when copper is in the anomalous limit at high frequencies and at lower temperatures. Again, the temperature dependences of $F_s$ and $X_s$ are probably more reliable than absolute values.

**b) Enclosed Perturbation: Sphere in a Maximum Electric Field.**
As discussed earlier, a small specimen can be placed at various positions inside the cavity, and the resulting frequency shift and change in width can be measured – and subsequently related to the conductivity components. As an example of the analysis, we discuss effects associated with the specimen placed in the maximum electric field.

The samples are assumed to be small (only adiabatic variations are considered) and we will therefore adopt the following convention: whenever the spatial dependence of a quantity is omitted, the quantity should be evaluated at the position of the sample (e.g., $\boldsymbol{P} = \boldsymbol{P}(\boldsymbol{r}_0)$ if the sample is located at the position $\boldsymbol{r} = \boldsymbol{r}_0$). If we put the sample in the antinode of the electric field ($\boldsymbol{H} = 0$), then

$$\boldsymbol{P} = \hat{\alpha}_e \boldsymbol{E}_s , \tag{4.66}$$

$$\frac{\Delta \hat{\omega}}{\omega} = -\frac{\hat{\alpha}_e}{4} V_s \frac{|\boldsymbol{E}_s|^2}{\langle W \rangle} = -4\pi \gamma \hat{\alpha}_e , \tag{4.67}$$

where $\gamma = \gamma_0 V_s / V_c$ with $V_s$ and $V_c$ the volume of the sample and of the cavity, respectively, and $\gamma_0$ is a constant that depends only on the resonance mode of the cavity

$$\gamma_0 = \frac{|\boldsymbol{E}_s|^2}{16\pi \langle Q \rangle} V_c = \frac{|\boldsymbol{E}_s|^2}{2\langle |\boldsymbol{E}|^2 \rangle} . \tag{4.68}$$

$E_s$ refers to the electric field at the position of the specimen (assumed to be constant for small samples) and where

$$\langle |\boldsymbol{E}|^2 \rangle = \frac{1}{V_c} \int_{V_c} |\boldsymbol{E}(\boldsymbol{r})|^2 dv . \tag{4.69}$$

The values of $\gamma_0$ are known for various cavities.

The absorption of electromagnetic waves by small specimens is proportional to the polarizability of the sample

$$\frac{\Delta\hat{\omega}}{\omega} = -4\pi\gamma\hat{\alpha}. \qquad (4.70)$$

The problem of determining $\Delta\hat{\omega}/\omega$ has then been reduced to finding the polarizability of an arbitrarily shaded sample. This requires one to solve the Helmholtz differential equation [4.2]:

$$\nabla^2 E(r) = 0 \quad r \text{ outside the sample}, \qquad (4.71)$$

$$\nabla^2 E(r) + \hat{k}^2 E(r) = 0 \quad r \text{ inside the sample}, \qquad (4.72)$$

where $\hat{k} = \omega/c_0\sqrt{\mu\varepsilon}$ in the complex wavevector inside the medium, subject to the appropriate boundary conditions (equivalent equations apply for the $H$ field).

The sample shape of lowest symmetry with solutions to (4.71, 72) which are independent of $r$, is the ellipsoid (4.72). In the case of an ellipsoid, the differential equation is called the Lamé equation, and the solutions are the ellipsoidal harmonics [4.22, 23]. The ellipsoidal coordinates are related to the Cartesian dimension through the elliptic functions of the first and second kind [4.24]. While solving the problem for an ellipsoid is appealing due to its generality, the solution is quite elaborate and cannot be expressed in terms of classical algebraic functions. Here, we will solve the problem for a sphere [4.10, 11, 25] will only quote the results for an ellipsoid. Other cases have been discussed by Klein et al.[4.3].

Depending on the ratio of the skin depth $\delta$ to sample size $a$, one can distinguish two limiting cases:

a) Depolarization Regime. $\hat{k}a \ll 1$. In this limit, the fields penetrate uniformly throughout the sample and one can effectively neglect the second term in (4.72). The resulting case reduces to a solution of Laplace's equation, just as in a static case, and under these conditions the sample is in the so-called depolarization regime. Here $\Delta' f$ refers to the shift with respect to a perfect conductor, for which the shift depth is zero.

b) Skin Depth Regime. $\hat{k}a \gg 1$. In this, the skin depth regime, $\hat{k}$ cannot be neglected, and one must solve the full set of Helmholtz equations. However, as we are not interested in the field distribution within the sample, we can use simple arguments to examine the form of the solution.

The effect of a sphere, placed in a maximum electric field has been calculated, [4.3] and the shift, together with the change of the width is displayed in Fig. 4.13. We note the following features. First, the absorption, $\Delta\Gamma$, peaks when $\varepsilon_1 = \varepsilon_2 + 1 - 1/n$, where $n = 1/3$ is the depolarization factor of the sphere. Second, the metallic shift ($\lim_{\hat{\sigma}\to\infty} \Delta\hat{\omega}/\omega_0$) is equal to $-\gamma/n$; and third the shift $\Delta' f/f_0$ changes sign and becomes negative in the skin depth regime. Here $\Delta' f$ refers to the shift with respect to a perfect conductor, for which the skin depth is zero.

In Fig. 4.13 we have replotted $\Delta\Gamma/2f_0$ and $|\Delta'f/f_0|$, normalized to the metallic shift $(\gamma/n)$, on a logarithmic scale to emphasize that, in the skin depth regime, $\Delta\Gamma/2f_0$ and $\Delta'f/f_0$ are both equal and proportional to the surface impedance $(\propto 1/\sqrt{\varepsilon_2})$. On this plot $\Delta'f/f_0$ changes sign.

We observe three independent regimes, each of which is characterized by a different power law dependence of the loss, $\Delta\Gamma$:

1) $\varepsilon_2 < \varepsilon_1 - 1 + 1/n$, this is the insulating side of the depolarization regime. In this range, $\Delta\Gamma/2f_0 \propto \sigma_1$ and the frequency shift saturates to a constant (which is proportional to $\varepsilon_1$).

2) $\varepsilon_1 - 1 + 1/n < \varepsilon_2 < 4\pi(c_0/\omega a)^2$, this is the metallic side of the depolarization regime, where the skin depth $(\delta = c_0/\omega\sqrt{2/\varepsilon_2})$ is still larger than the radius of the sphere $(a)$ but with the restriction $\varepsilon_2 > \varepsilon_1/n$. In this regime $\Delta\Gamma/2f_0 \propto 1/\sigma_1$, and the frequency shift goes asymptotically to the metallic shift from below.

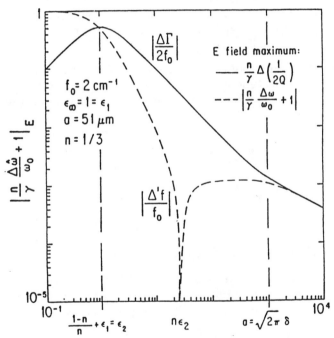

**Fig. 4.13.** The $\sigma_1$-dependence of the absolute value of $\Delta\Gamma$ and $\Delta'f$ for a sphere of radius $a = 5\,\mu$m, in the electric field maximum at $f = 2\,\mathrm{cm}^{-1}$ with $\varepsilon_1 = 1$. The $x$ scale is $n\varepsilon_2$, with $n = 1/3$ and $\varepsilon_2 = 4\pi\sigma_1/\omega$. We observe three distinct regimes, each characterized by a different pawer law dependence of the loss $\Delta\Gamma$. In the skin depth regime $\Delta\Gamma = -\Delta'f \propto \sqrt{1/\sigma_1}$, thus $\delta'\hat\omega/\omega_0$ is simply proportional to the surface impedance. The frequency shift $\Delta f/f_0$ is always negative, but the shift from a perfect conductor $\Delta'f/f_0 = \Delta f/f_0 + \gamma/n$ changes sign for $n\varepsilon_2 > 20$ (to become negative), and equals zero at the point of negative divergence [4.3]

3) $4\pi(c_0/wa)^2 < \varepsilon_2$, this is the skin depth regime. In this regime, $\Delta\Gamma/2f_0 \sim -\Delta' f/f_0 \propto 1\sqrt{\sigma_1}$ and one obtains,

$$\lim_{\sigma_1 \to \infty} \frac{\Delta\hat{w}}{w_0} \sim -\frac{\gamma}{n}\left(1 + \frac{ika}{2n\hat{e}}\right) = \gamma_n + \xi\hat{Z}_s,\qquad(4.73)$$

with $Z_s$ the complex surface impedance, and where the resonator constant $\xi$ is given by

$$\xi = \frac{-i\gamma}{n^2}\left(\frac{wa}{2c}\right) = -\frac{9}{2}i\gamma\frac{wa}{c}\qquad(4.74)$$

and the metallic shift is

$$\lim_{|\hat{\sigma}| \to \infty} \frac{\Delta\hat{w}}{w_0} = -\frac{\gamma}{n}.\qquad(4.75)$$

Solutions can also be obtained for a spheroid, both in an electric and in a magnetic field [4.3]. These expressions, in the limit when one axis is significantly larger than the other two axes, has been used both in the depolarization and in the skin depth regime to analyze the experimental results obtained on long, needle-shaped crystals.

c) **Ellipsoid in an Electric Field.** While rigorous solutions can be obtained in the case of a spherical specimen, expressions for the frequency shift and change of the quality factors have also been obtained for ellipsoids in certain limits.

Assuming that the ellipsoid has a depolarization factor $N$ in the limit where the electric field is uniform inside the specimen the following expressions apply:

$$\frac{\Delta w}{w} = \frac{\gamma}{N}\left(\frac{1 + N(\varepsilon'_1 - 1)}{[1 + N(\varepsilon_1 - 1)]^2 + (N\varepsilon_2)^2} - 1\right),\qquad(4.76)$$

$$\Delta\left(\frac{1}{Q}\right) = \frac{2\gamma}{N}\frac{N\varepsilon_2}{[1 + N(\varepsilon_1 - 1)]^2 + (N\varepsilon_2)^2},\qquad(4.77)$$

with $\varepsilon_1$ and $\varepsilon_2$ the real and imaginary parts of the dielectric constant. The above equations, when inverted, give the components of the complex dielectric constant (or complex conductivity) in terms of the experimentally accessible parameters.

The above equations break down in the skin depth regime as the electric fields become nonuniform inside the specimen. In this limit the shift and quality factor has been calculated under the assumption that $\sigma_1 \gg \sigma_2$; then the skin depth is given by $\sigma = c_0^2/2\pi w\sigma_1$, and the following relations apply:

$$\Delta\left(\frac{1}{Q}\right) = \frac{2\gamma}{N_3^2}\frac{9\pi\varepsilon_0 b}{2^4(2\varepsilon_2)^{1/2}},\qquad(4.78a)$$

$$\frac{\Delta\omega}{\omega} = -\Delta\left(\frac{1}{2Q}\right) - \frac{\gamma}{N_{\rm s}}, \tag{4.78b}$$

where, for a long needle, the depolarization factor is given by

$$N_{\rm s} = \frac{b^2}{a^2}\left[\ln\left(\frac{2a}{b}\right) - 1\right], \tag{4.79}$$

where $2a$ is the length and $2b$ the diameter of the ellipsoid [4.13, 14].

## 4.3 Experiments on Materials with Known Conductivity

The experimental methods, discussed in the previous section have been tested on materials with known conductivities which also were expected to be independent of the frequency in the millimeter with spectral range. The experiments were important in establishing the influence of various factors, such as sample alignment, temperature-induced changes in the waveguide and cavity characteristics, etc. These calibration experiments are summarized in this section.

### 4.3.1 Bridge Configuration Measurement

The bridge configuration, as described in Sect. 4.2 offers a straightforward measurement of the amplitude $A$ and phase $\phi$ from which the components of the complex conductivity can be extracted. There are, however, several factors which influence the scattering parameter $S$, and these factors have to be carefully examined before measurements on an unknown specimen can be conducted.

For the sample holder, as shown in Fig. 4.1, $l = 3\lambda_g/4$ only for one frequency. At an arbitrary frequency the short transforms into an impedance $i\tan(2\pi l/\lambda_{\rm g})$ in parallel with the sample impedance $Z_s$. Thus with the sample, the terminating impedance referred to the sample position is $Z = [1/iX_{\rm s} + 1/i\tan(2\pi l/\lambda_{\rm g})]^{-1}$, where we have assumed that the sample to be purely reactive ($Z_{\rm s} = iX_{\rm s}$), while without the sample, the impedance is a short at a distance $l$ from the sample position. Then using (4.22) with a transformed scattering parameter $S = -{\rm e}^{{\rm i}\phi_{\rm s}} \cdot {\rm e}^{{\rm i}4\pi l/\lambda_{\rm g}}$, the phase shift can be calculated

$$\phi_{\rm s} = -2\left[\cot^{-1}\left(\frac{1}{X_{\rm s}} + \frac{1}{\tan(2\pi l(\lambda_{\rm g})}\right) - \frac{2\pi l}{\lambda_{\rm g}}\right]. \tag{4.80}$$

For a given frequency $f$, $\phi_{\rm s}$ can thus be calculated using the expressions for $X_{\rm s}$, given in Sect. 4.2.

The measured phase shifts for quartz rods with different diameters are displayed in Fig. 4.14. For a $Ka$ band waveguide $a = 0.71\,{\rm cm}$ and $d/a = 0.20$, 0.14, and 0.053, respectively, for the three samples.

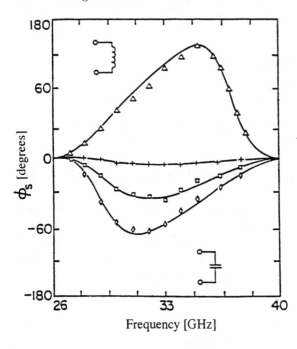

**Fig. 4.14.** Phase shift at room temperature as the function of frequency for the bridge configuration discussed in the text. *Triangles*: copper wire. *Squares, crosses* and *diamonds*: quartz rods. The equivalent impedances are also displayed

The measured phase shifts for a copper wire of diameter $d = 0.025$ cm are also plotted in Fig. 4.14. Using (4.50) for $X_s$, the calculated phase shift is also shown in Fig. 4.14 as a solid line. Again, the agreement is excellent.

The opposite signs of $\phi_s$ observed for the dielectric and the metal wire, as displayed in Fig. 4.14, demonstrate the ability of the bridge to directly determine the nature of reactance of an unknown sample, i.e., whether capacitive or inductive.

The material on which the temperature dependence of the conductivity was measured is TaSe$_3$. It has a conductivity of $2 \times 10^3\,(\Omega\text{cm})^{-1}$ at room temperature which increases by almost two orders of magnitude at $T = 10$ K. It is expected that at millimeter wave frequencies, the Hagen–Rubens limit applies, and $\sigma_1$ is independent of the frequency with $\sigma_1 \gg \sigma_2$.

The specimen is available in the form of thin fibers, and the measurements were performed primarily in the limit, $d < \delta$. This limit is ideal when sufficiently small samples are available because both the real and imaginary parts of the conductivity may be measured. From (4.57),

$$\sigma_1(T) + \mathrm{i}\sigma_2(T) = \frac{a}{Z_s(T) - \mathrm{i}X_\infty}, \tag{4.81}$$

where $Z_1(T)$ is determined at different temperatures from the measured $A_s$ and $\phi$, at each temperature as described in Sect. 4.2. The temperature dependence of $\sigma_1$ and $\sigma_2$ can be determined from $A_s$ and $\phi_s$, at each $T$ if $a$ and $X_\infty$ are known. Since $a$ and $X_\infty$ are purely geometric, being determined only

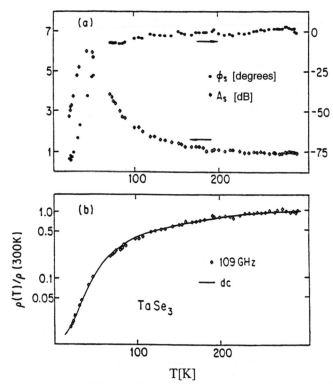

**Fig. 4.15.** (a) Phase shift $\phi_s$ and attenuation $A_s$ measured for a TaSe$_3$ sample as the function of temperature. (b) The temperature dependence of the resistivity of TaSe$_3$ as obtained from $A_s$ and $\phi_s$ as described in the text

by sample size and, hence, temperature-independent, it suffices to determine them at any convenient temperature.

The phase shift and attenuation due to the sample, $\phi_s$ and $A_s$, respectively, are shown in Fig. 4.15a.

The calculated resistivity, $\rho$ is displayed in Fig. 4.15b. The solid line is a four-probe dc resistivity on a sample from the same preparation batch. The dc resistivity changes by a factor of 50, and the 109 GHz resistivity follows it to within a scatter of $\pm 10\,\%$. At the lowest temperatures the classical skin depth $\delta$, is less than 1 μm and, therefore, $\delta \approx d$, but the agreement indicates that (4.81) still applies.

The measurement technique has also been extended for simultaneous measurements at several frequencies within one waveguide band [4.26].

**Cavity End-plate Measurements.** Cavity end-plate experiments have been conducted both on rectangular and cylindrical [4.27] cavities.

Extensive experiments on materials with known conductivity were conducted at room temperature using a rectangular cavity with one end-plate replaced by the material in question. As only part of the cavity is replaced, the

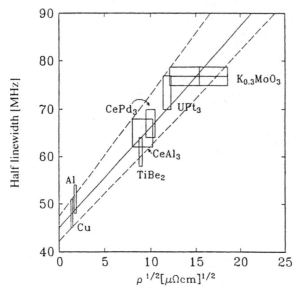

**Fig. 4.16.** Cavity half width measured with various materials as end-plates in a rectangular cavity. The *full line* represents the relation $R_s \simeq \rho_{dc}^{1/2}$, as expected in the Hagen–Rubens limit

change of the resonant frequency and width is related to $R_s$ and $Z_s$ through a geometrical factor $\gamma$. This factor represents the contribution of the replaced part to the frequency shift and inverse quality factor, which can be calculated. For the particular cavity in question one expects in the Hagen–Rubens limit that

$$\Delta f = \frac{\Delta(1/Q)}{2} = b_{th} \left(\omega\rho_{dc}\right)^{1/2} , \tag{4.82}$$

with $b_{th}$, which can be calculated for a particular geometry.

Both the shift and the change of the quality factor have been measured for materials with a moderate room temperature resistivity, and, as an example, the width of the resonance is displayed in Fig. 4.16 as a function of $\rho_{dc}$, measured on the same materials. The full line is (4.82). As can be seen on the figure, both parameters are described extremely well by the simple expression,

$$\Delta f = \frac{\Delta W}{2} = b_{exp}\rho_{dc} , \tag{4.83}$$

with $b_{exp}$ near to the calculated value. This, then, also gives evidence that these materials are, at room temperature, in the Hagen–Rubens regime, where $\sigma_1 \gg \sigma_2$, and $\sigma_2$ is independent of frequency.

The same method has been applied (by using a cylindrical cavity), to measure the temperature dependence of a material with a known (and moderate) conductivity, $TiSe_2$. Both $\rho_{dc}$ and the resistivity – evaluated by assuming again that the Hagen–Rubens limit applies – is displayed in Fig. 4.17. The

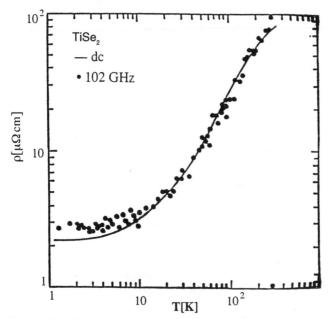

**Fig. 4.17.** Temperature dependence of $\rho_{dc}$ and of the resistivity at $100\,\text{GHz}$ of TiSe$_2$. It has been assumed that at $100\,\text{GHz}$ the Hagen–Rubens limit applies, and the relation between the components of the surface impedance and $\rho_{dc}$ is given by (4.84)

good agreement between the experiments performed at dc and at millimeter wave frequencies clearly demonstrates the usefulness of the cavity endplate replacement configuration, in measuring the conductivity of conducting solids.

### 4.3.2 Cavity Perturbation Measurements

As discussed earlier, various configurations can be used; the specimen can be positioned at different locations inside the cavity, either in a maximum electric field $\boldsymbol{E}$, or in a maximum magnetic field $\boldsymbol{H}$. Both configurations have been extensively employed, and here only a few examples will be discussed.

**a) Good Conductors: Surface Impedance Regime.** (TMTSF)$_2$PF$_6$ is a strongly anisotropic metal between room temperature and $12\,\text{K}$, at which temperature it undergoes a metal-insulator transition. The conductivity is highly anisotropic, with $\sigma$ in the direction of the $a$-axis well exceeding the conductivity in the other two directions. The normal state conductivity at room temperature is large ($\rho_{dc} \approx 10^{-3}\,\Omega\,\text{cm}$ [4.28] with a room temperature $60\,\text{GHz}$ skin depth of approximately $5\,\mu\text{m}$. As the sample dimensions exceed the skin depth, one expects the surface impedance formalism to be appropriate at all temperatures above the metal-insulator transition.

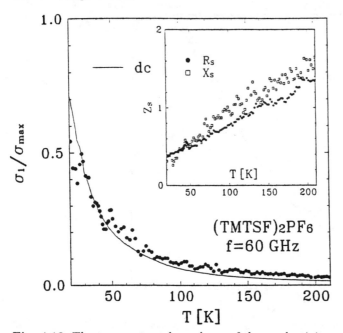

**Fig. 4.18.** The temperature dependence of the conductivity $\sigma_1$ of $(TMTSF)_2PF_6$ measured with the magnetic field perpendicular to the chain direction. The *solid line* is the dc resistivity. In the *inset* we display the normal state surface resistance $R_s$ and surface reactance $X_s$, and as expected for a normal metal, $R_s = X_s$ over a broad temperature range. Both the dc and the high-frequency conductivity is normalized to the room temperature value

In the inset of Fig. 4.18 we display the temperature dependence of both the surface resistance $R_s$ and the surface reactance $X_s$, up to 200 K. The good overall agreement between the two quantities clearly indicates that the metallic behavior ($\sigma_1 \gg \sigma_2$) persists up to at least 200 K, as in this limit:

$$R_s = X_s = \left(\frac{\mu_0 \omega \rho_{dc}}{2}\right)^{1/2}.$$
(4.84)

From the measured values of $R_s$ and $X_s$ one can determine the components of the conductivity,

$$\sigma_1 = \frac{\omega}{2\pi} \frac{R_s X_s}{(R_s^2 + X_s^2)^2},$$
(4.85)

and

$$\sigma_2 = \frac{\omega}{4\pi} \frac{(X_s^2 - R_s^2)}{(R_s^2 + X_s^2)^2}.$$
(4.86)

The real part, $\sigma_1$ is displayed, together with the dc conductivity in Fig. 4.18; within experimental error the imaginary part, $\sigma_2$ is zero. Again, we find that $\sigma_1$ is independent of frequency; this is expected for a material with a moderate conductivity.

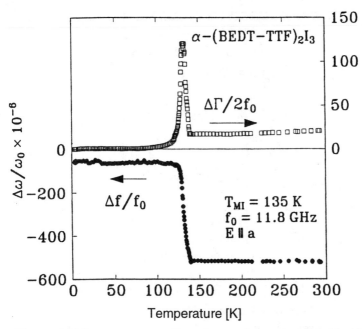

**Fig. 4.19.** The temperature dependence of the relative half width $\Delta\Gamma/2f_0$ and change of the resonance frequency $\Delta f/f_0$ for $\alpha-$ (BEDT–TTF)$_2$I$_3$ in 11.8 GHz cavity. Near the phase transition at $T_{MI} = 135$ K the sample response changes from the insulating to the metallic regime as seen in the large change of the frequency shift $\Delta f$ and the peak in the bandwidth $\Delta\Gamma$

**b) Poor Conductors: Depolarization Regime.** At room temperature the $\alpha$-phase of (BEDT-TTF)$_2$I$_3$ shows a two-dimensional dc metallic conductivity between 60 and 250 $\Omega$ cm with $\sigma_b/\sigma_a \approx 2$ and $\sigma_a/\sigma_c \approx 2\,000$. At $T_{MI} = 135$ K it undergoes a metal-insulator phase transition and the dc electric conductivity is seen to drop sharply by several orders of magnitude [4.29].

In Fig. 4.19 we display the temperature dependence of the change in both the width $\Delta\Gamma/2f_0$ and resonance frequency $\Delta f/f_0$. The phase transition at $T_{MI} \approx 135$ K causes a large, rapid change in the resonance frequency. Additionally, the bandwidth change also increases sharpy just above $T_{MI}$, but in this case a peak is observed just below $T_{MI}$ before a decrease to lower temperatures.

The microwave conductivity is displayed in Fig. 4.20 as a function of temperature together with the dc results, with the insert displaying the dielectric constant.

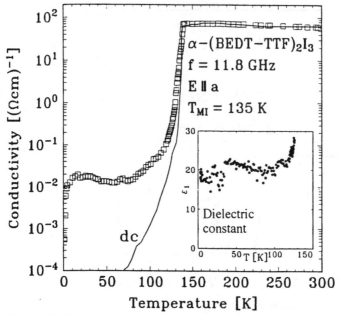

**Fig. 4.20.** The microwave conductivity along the axis is displayed together with the dc conductivity for $\alpha-$ (BEDT–TTF)$_2$I$_3$. Below the metal-insulator phase transition the high-frequency conductivity develops a plateau while the dc values continue to decrease. The inset shows the temperature dependence of the dielectric constant at 12 GHz

## 4.4 Experiments in Correlated Metals

The name, correlated metals, refers to metallic systems where electron-electron or electron-photon interactions play an important role. These interactions may lead to important modifications of the thermodynamics and of the electrodynamics of the metallic state. They also can lead to phase transitions to states where a particular symmetry is broken, hence the name of broken symmetry ground states.

These interactions also lead to reduced energy scales, and therefore the spectroscopic range we have discussed plays an important role. The topics covered in this section include materials in their metallic state, and two types of broken symmetry states: superconductors and density waves.

### 4.4.1 Heavy Fermion Materials

In several intermetallic compounds which very large specific heat coefficients and low temperature Pauli susceptibilities are observed [4.30–32].

These unusual properties are associated with a narrow resonance appearing at the Fermi level in the low temperature "coherent" regime and above $T^*$

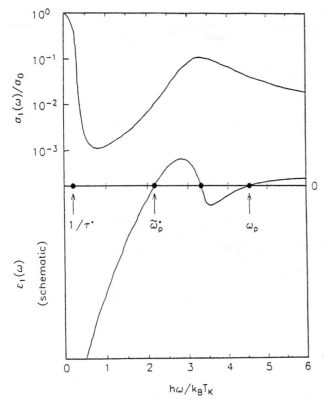

**Fig. 4.21.** The calculated dynamical conductivity and dielectric constant for Heavy Fermions [4.37]. The zero-crossing at $\tilde{\omega}_p^*$ corresponds to the renormalized plasma frequency, and at $\omega_P$ to the renormalized plasma frequency. In the Figure, $\sigma_r$ refers to the real part of the conductivity and $\varepsilon_r$ to the real part of the dielectric constant

this feature disappears [4.33, 34]. Various models based on the so-called Anderson Hamiltonian lead to a so-called "hybridization gap" of energy  and to a narrow-many-body resonance both centered on $B^*$ above the Fermi surface. A temperature and frequency-dependent scattering rate is also predicted, and the resulting conductivity $\sigma_r(\omega)$ and dielectric constant $\varepsilon_2(\omega)$ are schematically displayed in Fig. 4.21 [4.35–38].

The essential features of Fig. 4.21 include the following: First, there is a narrow absorption at zero frequency that is describable by a renormalized Drude behavior of the form

$$\sigma_r(\omega) = \frac{ne^2\tau^*}{m^*}\frac{1}{1+(\omega\tau^*)^2}, \tag{4.87}$$

where $n$ is the combined total number of conjuction and $f$ electrons (i.e., $n = n_c + n_f$). Notice that the dc conductivity from (4.87), $\sigma_{cd} = ne^2r^*/m^\delta$, is unrenormalized because of (6.4). Above the Drude roll-off at $1/\tau^*$, there

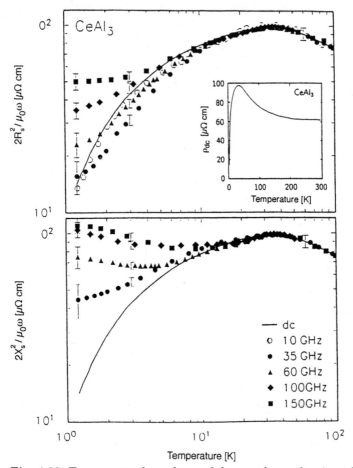

**Fig. 4.22.** Temperature dependence of the complex surface impedance, and the dc resistivity of GeAl₃. The choice of the particular representations of $R_{\mathrm{s}}$ and $X_{\mathrm{s}}$ is made to show the frequency-independent behavior at high temperatures. The *full line* is the dc resistivity

are two plasma frequencies associated with zero-crossings of the dielectric function. The low frequency one is a renormalized plasma frequency given by

$$\omega_{p1}^* = 6^{1/2} \left( 2 + \frac{n_{\mathrm{c}}}{n_{\mathrm{f}}} \right)^{1/2} T^* . \tag{4.88}$$

$T^*$ is the renormalized Fermi temperature.

Typical experimental results, obtained on the material CeAl₃ are displayed in Fig. 4.22 [4.39]; where we have plotted the parameters $2R_{\mathrm{s}}^2/\mu_0\omega$ and $2X_{\mathrm{s}}^2/\mu_0\omega$ for the following reasons. In the normal skin-depth regime the surface impedance and the complex conductivity are related as

$$Z_s = R_s - iX_s = \sqrt{\frac{-i\mu_0\omega}{\sigma_1 - i\sigma_2}} \, . \tag{4.89}$$

In the high-temperature range the scattering rate $(1/\tau)$ is significantly larger than the millimeter-wave frequencies, and $\sigma_2$ is negligible compared to $\sigma_1$. Consequently, the following approximations of (4.87) (the so-called Hagen–Rubens limit) are valid at these temperatures

$$Z_s = \sqrt{\frac{-i\mu_0\omega}{\sigma_{dc}}} \, , \tag{4.90}$$

$$R_s = X_s = \left(\frac{\rho_{dc}\mu_0\omega}{2}\right)^{1/2} \, , \tag{4.91}$$

where $\rho_{dc}$ is the dc resistivity. As is evident from Fig. 4.22, this relation holds well at high temperatures, and the conductivity is independent of frequency in the millimeter wave spectral range.

For the purpose of the Kramers–Kronig analysis, the microwave and optical measurements were combined in order to obtain the absorptivity over a broad spectral range. The absorptivity and surface impedance are related according to the equation, see (4.19),

$$A = \frac{4R_s}{Z_0} \frac{1}{1 + \frac{2R_s}{Z_0} + \frac{R_s^2 + X_s^2}{Z_0^2}} \, . \tag{4.92}$$

The last term in the denominator of this expression is negligible compared to the first term since $R_s, X_s \ll Z_0$ in the submillimeter to microwave spectral range, for a highly conducting material such as $CeAl_3$ and (4.20) applies. Combining these results together with $A = 1 - R$, where $R$ is the optical reflectance from 15 to $10^5$ cm$^{-1}$, we obtain the absorptivity from microwave frequencies up to 12 eV, shown in Fig. 4.23. The absolute absorptivity at 15 cm$^{-1}$ was normalized to the 10 K Hagen-Rubens value. The $\pm 15\,\%$ error in the absorptivity at 15 cm$^{-1}$ results from a 0.5 % uncertainty of the reflectivity over the FIR spectral range.

Below about 10 K, both $R_s$ and $X_s$ deviate from what would follow from a frequency independent conductivity, suggesting a narrow, frequency dependent feature in the millimeter wave spectral range. Using the relation (4.15b), $\sigma_1$ and $\sigma_2$ can be extracted using the measured $R_s$ and $X_s$ values and these are displayed in Fig. 4.24. As expected, $\sigma_1$ is independent of frequency at $T = 10$ K and develops a strong frequency dependence at $T = 1.2$ K the so-called coherent regime.

The optical conductivity at 10 K, reflects the Drude behavior with a relaxation rate $1/\tau = 1\,000$ cm$^{-1}$ and $\hbar\omega_p = 4$ eV. The value of the plasma frequency is in agreement with the evaluation of $\omega_p$ as inferred from the total spectral weight (below). The agreement between the extrapolation of $\sigma_1(\omega)$ for $\omega \to 0$ obtained from the Kramers–Kronig analysis and $\sigma_{dc}$ is excellent. At 1.2 K we observe a narrow resonance in $\sigma_1(\omega)$ and the data joints the

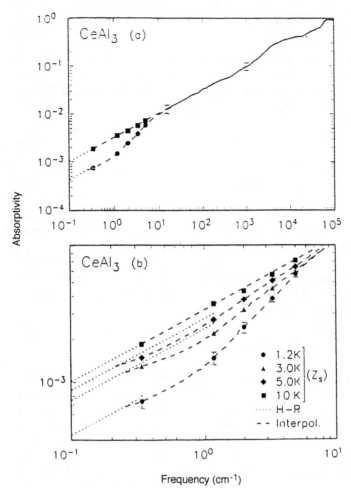

**Fig. 4.23.** (a) Frequency dependence of the absorptivity $A = 1 - R$ of CeAl3 at 1.2 and 10 K. Both optical and microwave data are displayed in the figure. (b) Temperature-variation of $A$ in 0.1–10 cm$^{-1}$ range. Notice the dip in the absorptivity below the Hagen–Rubens line at 3 K

10 K data at approximately 3 cm$^{-1}$. Again, $\sigma_1(\omega)$ from the Kramers–Kronig analysis and $\sigma_{dc}$ at 1.2 K agree quite well. The error in the absolute value of $\sigma_1(\omega \to 0, 1.2$ K$)$ is $\approx 12\%$ reflecting the inaccuracies in the absolute value of the reflectance at low frequencies.

It is evident from the figure, that an extremely narrow resonance develops at low temperatures. In order to analyze this feature, and to extract quantities which are related to the enhanced effective mass, one can evaluate the following integrals from our experimentally obtained $\sigma_1(\omega)$:

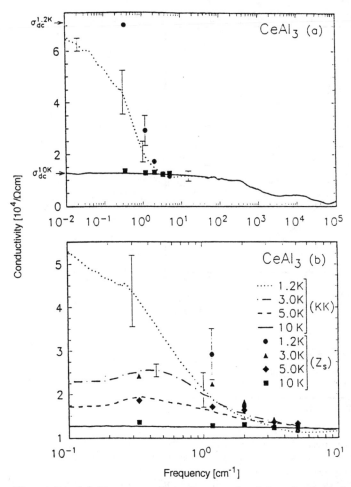

**Fig. 4.24.** (a) Frequency-dependent conductivity obtained from the Kramers–Kronig analysis of the data (KK) in Fig. 4.23, and directly from the surface impedance measurements $Z_s$. (b) A close-up of the dynamical conductivity showing the development of the low-frequency resonance at low temperatures. The 10 GHz point from the surface resistance date is equal to $(2R_s^2/\mu_0\omega)^{-1}$ giving evidence that the conductivity is independent of frequency up to 10 GHz

$$I_1 = \int_0^{\omega_c} \sigma_1(\omega)d\omega = \frac{\pi ne^2}{2m^*} , \qquad (4.93)$$

and

$$I_2 = \int_0^{\infty} \sigma_1(\omega)d\omega = \frac{\pi ne^2}{2m} , \qquad (4.94)$$

where $\omega_c = 3\,\mathrm{cm}^{-1}$ is the frequency above which the data at 10 K and at 1.2 K are indistinguishable. From the resulting ratio $I_2/I_1$ we evaluate

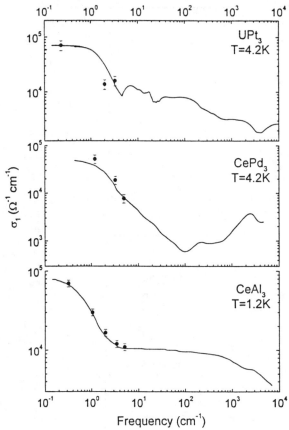

**Fig. 4.25.** The real part of the conductivity vs frequency for CePd₃, UPt₃ and CeAl₃. For all three samples the temperature was 1.2 K. In the case of CeAl₃, all the values were directly measured. For CePd₃ and UPt₃, certain assumptions were utilized in the analysis (4.39). The *solid lines* are guides to the eye, and the very narrow zer frequency resonance is a manifestation of the strong electron-electron correlations

$m^*/m_b = 450 \pm 50$, a value comparable with that estimated from the thermodynamic quantities. It is increasingly more difficult to estimate the temperature dependence of $m^*/m_b$ since the spectral weight due to the narrow resonance decreases dramatically and the ratio of $I_1/I_2$ is harder to estimate as the temperature increases.

The narrow resonance observed in CeAl₃ has also been found in other highly correlated metallic systems and in Fig. 4.25 $\sigma_1(\omega)$ is displayed in three well-studied heavy fermions materials.

**Superconductors.** The electrodynamics of the superconducting state has been discussed at length in several books and reviews [4.43]. The response depends on factors like the penetration depth $\lambda$, mean free path $l$ and coherence length $\xi$. The frequency-dependent conductivity has been calculated in the various limits, and strong coupling effects have also been examined.

Here, the variety of observations will not be summarized, but we will focus on one particular aspect of the field: the observation of the complex conductivity at frequencies well below the gap frequency $\omega_g = \Delta/\hbar$; the experiments, conducted on Nb ($T_c = 9.3\,\mathrm{K}$ and $\Delta$ 16 K) will be contrasted with the theory.

A complex conductivity was calculated by *Mattis* and *Bardeen* [4.44]. According to this calculation, based on the BCS theory, $\sigma_1(T)$ shows a peak just below $T_c$ at low frequency. The height of the peak is given approximately by

$$\frac{\sigma_1^*}{\sigma_n} \sim \log\left(\frac{2\Delta(0)}{\hbar\omega}\right), \tag{4.95}$$

where $\sigma_1^*/\sigma_n \equiv (\sigma_1/\sigma_n)\mathrm{max}$. The peak completely disappears for $\hbar\omega \gtrsim \Delta/5$ (at frequencies $2\Delta$). The width of the peak is slightly frequency-dependent. $\sigma_1(T)/\sigma_n$ can be approximated by the following algebraic formula:

$$\frac{\sigma_1(T)}{\sigma_n} \sim 2f[\Delta(T)]\left[1 + \frac{\Delta(T)}{k_B T}\{1 - f[\Delta(T)]\}\ln\left(\frac{2\Delta(T)}{\hbar\omega}\right)\right], \tag{4.96}$$

and $\sigma_2/\sigma_n$ is related to the gap parameter through the expression:

$$\frac{\sigma_1(T)}{\sigma_n} \simeq \frac{\pi\Delta(T)}{\hbar\omega}\tanh\left(\frac{\Delta(T)}{2k_B T}\right). \tag{4.97}$$

Using the relation between the surface impedance $Z_s$ and complex conductivity $\sigma$ (4.15), the components $R_s$ and $X_s$ can also be calculated. Having established the relation between $Z_s$ and $\sigma$, the components of the measured surface impedance can then be used to evaluate the components of the complex conductivity. In terms of the complex conductivity these parameters are given by inverting (4.15) as

$$R_s = \sqrt{\frac{\omega}{4\pi}}\frac{1}{\sqrt{\sigma_1^2 + \sigma_2^2}}\sqrt{-\sigma_2 + \sqrt{\sigma_1^2 + \sigma_2^2}}, \tag{4.98a}$$

and

$$X_s = \sqrt{\frac{\omega}{4\pi}}\frac{1}{\sqrt{\sigma_1^2 + \sigma_2^2}}\sqrt{-\sigma_2 + \sqrt{\sigma_1^2 + \sigma_2^2}}. \tag{4.98b}$$

In the superconducting phase, where $\sigma_2 \gg \sigma_1$,

$$X_s = \frac{\omega}{c}\lambda_0, \tag{4.99}$$

and $\lambda_0$ is the penetration depth,

$$\lambda_0 = \frac{c}{\omega_p}, \tag{4.100}$$

and $\omega_p = (4\pi n e^2/m)^{1/2}$ is the plasma frequency. (4.100) applies in the clean limit, in the dirty limit spectral weight arguments can be developed to relate

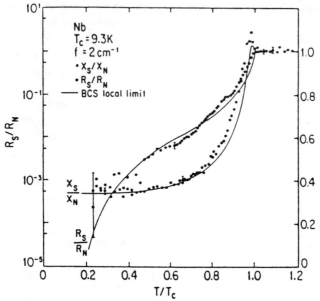

**Fig. 4.26.** Temperature dependence of the surface resistance $R_s$ and the surface reactance $X_s$ of Nb. The surface resistance is displayed on a logarithmic scale while the surface reactance is plotted on a linear scale. The *solid line* is the computed $Z_s$ from the conductivity calculated by *Mattis* and *Bardeen* [4.44], assuming that the measurement is in the local limit

$\lambda_0$ to the optical parameters [4.43]. At low temperatures, where $\Delta(T)$ has an exponential temperature dependence, (4.97) after expansion gives

$$\frac{\sigma_2}{\sigma_1} \sim \left(\frac{\omega}{\Delta}\right)^2 \exp\left(\frac{\Delta}{k_B T}\right) , \qquad (4.101)$$

leading to

$$\frac{R_s}{R_N} = A \frac{\Delta}{T} \left(\frac{\omega}{\Delta}\right)^2 \ln\left(\frac{\Delta}{\hbar \omega}\right) \exp\left(-\frac{\Delta}{k_B T}\right) , \qquad (4.102)$$

with $A$ a constant.

In Fig. 4.26 the surface reactance and surface resistance measured in Nb is displayed. Both parameters are normalized to the value obtained in the normal regime, above $T_c$. As discussed earlier, $X_s$ can be evaluated only up to a numerical (temperature-independent) constant, and it was assumed that in the normal state $R_s = X_s$ in agreement with the Hagen–Rubens limit. From the surface resistance one finds a normal state conductivity $\sigma_n = 0.85 \times 10^6 \ \Omega^{-1} \text{cm}^{-1}$.

In the superconducting state, the temperature dependence of the penetration depth is given by

$$\frac{X_s(T)}{X_n} = \frac{2\lambda_0(T)}{\delta} . \qquad (4.103)$$

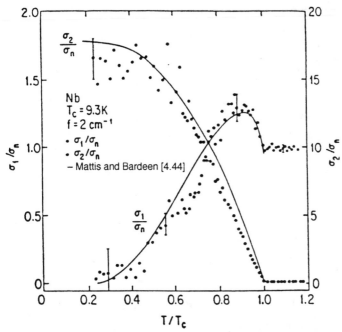

**Fig. 4.27.** Temperature dependence of the complex conductivity of Nb evaluated from surface impedance measurement. The *solid line* is the *Mattis* and *Bardeen* [4.44] prediction [4.46]

At $T = 0$ one obtains $\lambda_0 = 440\,\text{Å}$ in good agreement with the penetration depth measured by other methods (*Klein* [4.45]). The temperature dependence of both $R_s$ and $X_s$ can be well described by the expressions developed by *Mattis* and *Bardeen* [4.44], and the full lines are the calculated values assuming that the weak coupling BCS limit applies.

Using the measured $R_s$ and $X_s$ values together with (4.98) the temperature dependence of the complex conductivity can be established, and both $\sigma_1(T)$ and $\sigma_2(T)$ are displayed in Fig. 4.27. The full lines are again the conductivity components calculated using the theory of *Mattis* and *Bardeen* [4.44]. The maximum observed for $\sigma_1$ below $T_c$ is due to the so-called case II coherence factors [3.43, 47].

Using $\sigma_2(T)$ one can also evaluate the temperature dependence of the single particle gap by applying (4.97). $\Delta(T)$ obtained is displayed on Fig. 4.28, together with the temperature dependence which follows from the BCS weak coupling limit.

Similar experiments have been conducted in Pb, where the anomalous limit applies [4.3] and also strong coupling corrections are important. Experiments similar to those shown for Nb have been recently conducted on a variety of novel superconductors, and some of the results have been reviewed by *Dressel* et al. [4.27].

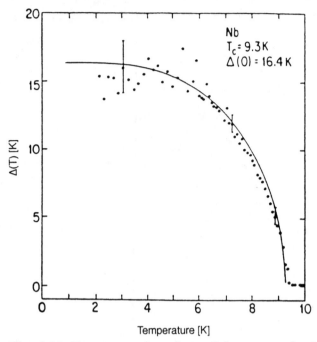

**Fig. 4.28.** Temperature dependence of the superconducting gap of NB. The *full line* is the calculated temperature dependence of the gap, using the BCS theory [4.46]

### 4.4.2 Density Waves

Density waves are condensates formed instead of electron-electron pairs (as in the case of superconductors) by electron-hole pairs. The thermodynamics of the states is the same as that of superconductors, and there is a second-order phase transiton at temperature $T_{DW}$ which leads to these ground states. Below $T_{DW}$ there is a well defined single particle gap, and it is related to the transition temperature by the well-known BCS relation $2\Delta = 3.52k_B T_{DW}$. The temperature dependence of $\Delta$ is also given by the BCS expression.

The name Charge density Wave (CDW) and Spin Density Wave (SDW) refers to condensates when the charge density or spin density has a periodic modulation. In the former case the charge density is given by

$$\Delta\rho = \rho_1 \cos(2k_F + \phi'),$$

where $\rho_1$ is the amplitude, $\phi'$ is the phase of the charge density wave, and $k_F$ is the Fermi wavevector; the spin density has the spatial dependence,

$$\Delta s = S_1 \cos(2k_F r + \phi'),$$

in the SDW phase. Here $\rho_1$ and $S_1$ depend on the parameters like single particle bandwidth, electron-phonon and electron-electron interactions.

The dynamics of the collective mode is described in terms of the time dependent order parameter [4.48]. As the order parameter has two degrees of freedom (amplitude and phase), both amplitude and phase fluctuations will occur.

The excitations, which correspond to local modulation of the phase $\phi$ with a period $q$ are called phasons.

In the presence of an applied electric field, the equation of motion for the excitations is

$$\frac{d^2\phi}{dt^2} + v_F^2 \frac{m}{m^*} \frac{d^2\phi}{dx^2} = \frac{2k_F eE}{m^*} , \tag{4.104}$$

and the frequency dependent conductivity,

$$\sigma(\omega) = \frac{j(\omega)}{E(\omega)} = \frac{m}{m^*} \frac{i\omega_P^2}{4\pi(\omega + i\delta)} , \tag{4.105}$$

with the real part,

$$\text{Re}\{\sigma(\omega)\} = \frac{ne^2}{m^*}\delta(\omega). \tag{4.106}$$

The $\text{Re}\{\sigma(\omega)\}$ peak at zero frequency, a feature, reminiscent of superconductivity.

Experiments performed at low dc electric fields do not show evidence for the collective mode. This is because impurities, lattice imperfections, etc. interact with the CDW and they pin the phase to the underlying lattice. Representing this by an average pinning frequency $\omega_P$, and also including a phenomenological damping constant $\Gamma$, the phenomenological equation of $q = 0$ motion for the pinned mode is that of a harmonic oscillator,

$$m^*\ddot{x} + \Gamma\dot{x} + Kx = eEe^{i\omega t} , \tag{4.107}$$

where $K = \omega_P^2 m^*$. This leads to an ac conductivity,

$$\sigma(\omega) = \frac{ne^2}{i\omega m^*} \frac{\omega^2}{\omega_P^2 - \omega^2 - i\omega\Gamma} . \tag{4.108}$$

The low frequency dielectric constant $\varepsilon(\omega \to 0)$ is given by

$$\varepsilon(\omega \to 0) = \frac{4\pi ne^2}{m^*\omega_P^2}. \tag{4.109}$$

One also recovers single particle excitation across the charge or spin density wave gap, and the expected $\omega$-dependent response for $\omega_0 \ll \Delta$ is shown in Fig. 4.29.

Experiments have been conducted on various materials with a charge density wave ground state, and typical results, obtained on the compound $(TaSe_4)_2I$, are displayed in Fig. 4.30. $\sigma_1$ and $\varepsilon_1$ are obtained from the measured attenuation $A$, and phase shift $\phi$, the quantities obtained using a bridge configuration as discussed in Sect. 4.4. While the dc conductivity rapidly decreases below the phase transition at $T_{CDW} = 260\,\text{K}$, the conductivity at

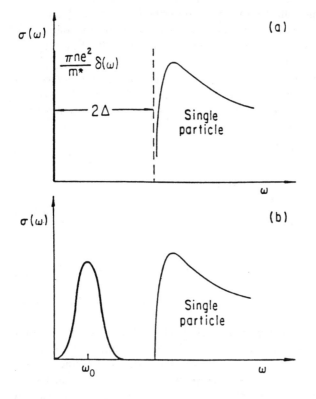

Fig. 4.29. Expected frequency dependent response for density wave condensates, with (a) and without pinning (b) of the collective mode. The resonance at $\omega_0$ is due to the pinned collective mode, executing Oscillations about the equilibrium position

millimeter wave frequencies remains high, suggesting the development of a resonance.

Similar results were obtained at other compounds, and $\sigma(\omega)$ evaluated in broad spectral range for various materials is shown in Fig. 4.31 on the figure.

The low-frequency dielectric constant is enormous, in general, of the order of $10^7$ or more, and this is the consequence of the large oscillator strength which occurs at low frequencies. The zero-frequency dielectric constant is given by

$$\varepsilon_1(\omega \to 0) = \frac{4\pi e^2}{m^*} \int_0^\infty \frac{\sigma_1(\omega)}{\omega^2} d\omega, \tag{4.110}$$

which, for a narrow resonance reduces to $\varepsilon_1 \sim 4\pi n e^2/\omega_{\mathrm{P}}^2$.

Similar experimental results were obtained in the SDW phase of the model compound $(TMTSF)_2PF_6$ where the collective mode has also been identified by experiments in the millimeter wave spectral range [4.51, 52]. In these materials the effective mass is not enhanced and consequently, all the spectral weight which is removed from the conductivity should be in the collective mode. The experimental results are displayed in Fig. 4.32. Above the SDW transition, which occurs at $T_{\mathrm{SDW}} = 11.5\,\mathrm{K}$, the low frequency response is

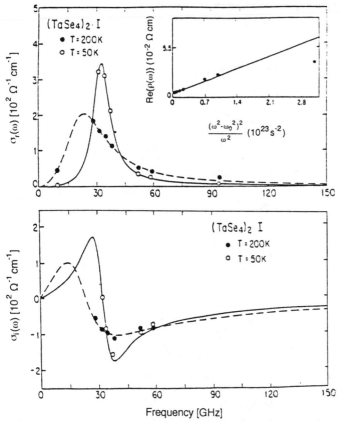

**Fig. 4.30.** Frequency dependence of the real part, $\sigma_r$ (*upper part*) and imaginary part $\sigma_i$ (*lower part*) of the conductivity in $(TaSe_4)_2I$ in the charge density wave state. The *solid lines* are fits to the harmonic oscillator expression (4.107) [4.48–50]

Drude-like, with an additional (probably) interband transition which occurs around $100\,cm^{-1}$. Below $T_{SDW}$, the narrow resonance around $4\,GHz$ is due to the pinned collective mode, as established by experiments conducted on alloys [4.53].

## 4.5 Conclusion

Millimeter wave spectroscopy has evolved over the years into an important research tool for investigating the dynamics of solids. This is due to the development and availability of components in this spectral range, which, nevertheless, remains expensive when compared to research tools utilized at lower or higher frequencies. The importance of the technique when studying highly correlated solids, both in the Fermi-liquid and broken-symmetry

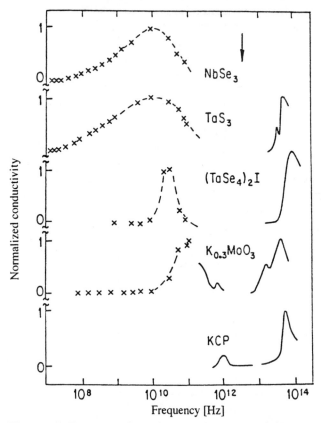

**Fig. 4.31.** Frequency-dependent conductivity $\sigma_1(\omega)$ for various compounds with a charge density wave ground state. Both excitations across the single particle gaps (*full lines* in the infrared spectral range) and the collective mode resonances (*dashed lines* in the millimeter wave spectral range) are displayed [4.48]

states, is a consequence of the dramatically reduced energy scales characterizing the correlations and the dynamic response. Several techniques, using quasi-optical arrangements and nonresonant cavity waveguide configurations, are frequently applied to measure the properties of relatively highly conducting solids. Invariably, they measure one component of the response, such as reflection or absorption, and an evaluation of the complex conductivity requires either an assumption regarding $\sigma(\omega)$ (such as a Drude or Lorentz model), or a Kramers–Kronig analysis. Both approaches have obvious disadvantages. The techniques discussed here measure two components: the attenuation and phase shift in case of a bridge configuration or the quality factor and resonant frequency in the case of a cavity measurement, and the two components of the conductivity can be evaluated without any further assumptions. This represents the clearest advantage of the technique. With resonant cavities, the sensitivity is an added important advantage, particularly for highly conduct-

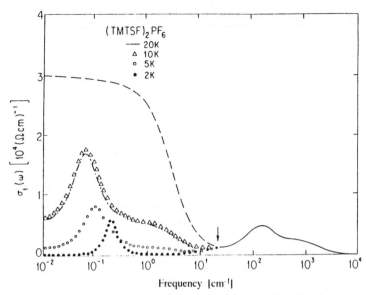

**Fig. 4.32.** Frequency dependent conductivity $\sigma_1(\omega)$ for (tetramethyltetraselen-afulvalene)$_2$ PF$_6$, (TMTSF)$_2$PF$_6$ in the metallic ($T > T_{\mathrm{SDW}} = 11.5\,\mathrm{K}$) and in the spin density wave state ($T < 11.5$). The Drude response at $T = 20\,\mathrm{K}$ has a reduced spectral weight indicating a somewhat enhanced bandmass $m_\mathrm{b}$; the narrow resonances in the SDW state are due to the phason excitations. The data have been obtained by combining surface impedance and reflectivity data, and by subsequent Kramers–Kronig analysis [4.51]

ing samples. Broadly speaking, for a good metal, the deviation of the optical reflectivity from perfect reflection, $1 - R$, is proportional to $R_\mathrm{s}/Z^0$ where $Z^0$ is the impedance of free space. The ratio $R_\mathrm{s}/Z^0$ is measurable to an accuracy approaching $10^{-5}$, and this exceeds the performance of optical reflectivity measurements by about three orders of magnitude. The clear disadvantage of the method is the fact that measurements can be conducted at only selected frequencies, requiring a change in the configuration, for example, the cavity size, for each experimental point. The development of continuous-frequency measurements (by employing a transmission bridge configuration or tunable cavity measurement) would be a significant advancement.

The examples discussed in this review involve energy scales significantly smaller than typical single particle energies such as the unrenormalized relaxation rate $1/\tau$. For uncorrelated metals $1/\tau$ is of the order of $10^{-15}\,\mathrm{s}$, resulting in a corresponding frequency that lies in the optical spectral range. The single particle gaps, which characterize the broken-symmetry ground states, are related to the transition temperatures $T_\mathrm{c}$. In the weak coupling limit (for isotropic gaps), they are given by $2\Delta = 3.5k_\mathrm{B}T_\mathrm{c}$ and for transition temperatures of $10\,\mathrm{K}$ or less, the single particle gaps lie in the millimeter wave spectral range. Concerning the renormalized single particle response in the Fermi-liquid state or the collective mode response for the charge and spin density wave states, the peak conductivity approaches that of the uncorrelated metals. Consequently, the sum rule,

$$\int \sigma_1(\omega)d\omega = \frac{\pi n e^2}{2m^*},$$  (4.111)

implies that the spectral width of the response is approximately

$$W \approx \frac{ne^2}{m^*\sigma_{dc}},$$

making $W$ inversely proportional to the effective mass $m^*$. For large effective masses, like those which characterize some heavy-fermion materials or charge density wave systems, the important frequency dependences occur again at millimeter wave frequencies. From a Drude or Lorentz approach, this leads to relaxation times $\tau^*$ which are enhanced over the relaxation times characterizing the transport in the absence of correlations.

It is expected that millimeter wave spectroscopy will play an important role in the investigation of the dynamical response in other fields of condensed matter physics where reduced energy scales are important. Disorder induced metal-insulator transitions, the exploration of exotic ground states possibly existing in the superconducting state of some heavy fermion or organic superconductors, the quantized Hall effect, and the magnetic field-induced spin density wave states are a few obvious examples. Further developments in the technique enabling experiments at lower temperatures and high magnetic fields will be required to explore the unusual dynamics associated with these types of correlated solids.

# References

4.1    F. Wooten: *Optical Properties of Solids* (Academic, New York 1972)
4.2    J.D. Jackson: *Classical Electrodynamics*, 2nd edn. (Wiley, New York 1975)
4.3    O. Klein, S. Donovan, M. Dressel, G. Grüner: Int'l J. Infrared Millimeter Waves **14**, 2423 (1993)
4.4    J. Müüller: Hochfrequenztechnik und Elektroakustik **54**, 157 (1939)
4.5    J.C. Slater: Rev. Mod. Phys. **18**, 441 (1946)
4.6    J.C. Slater: *Microwave Electronics* (Van Nostrand, New York 1950)
4.7    C.G. Montgomery: *Technique of Microwave Measurements*, MIT Rad. Lab. Ser., Vol. 11 (McGraw, New York 1947)
4.8    S. Ramo, J.R. Whinnery, T. Van Duzer: *Fields and Waves in Communication Electronics* (Wiley, New York 1984)
4.9    S. Donovan, O. Klein, M. Dressel, K. Holczer, G. Grüüner: Int'l J. Infrared Millimeter Waves **14**, 2459 (1993)
4.10    K.S. Champlin, R.R. Krongarad: IRE Trans. MIT-**9**, 545 (1961)
4.11    M.E. Brodwin, M.K. Parson: J. Appl. Phys. **36**, 494 (1965)
4.12    L.I. Buranov, I.F. Schegolev: Pribory i Tekhnika Eksperimenta **2**, 171 (1971)

4.13   M. Cohen, S.K. Khanan, W.J. Gunning, A.G. Garito, A.J. Heeger: Solid State Commun. **17**, 367 (1975)

4.14   N.P. Ong: J. Appl. Phys. **48**, 2935 (1977)

4.15   J. Schwinger, D.S. Saxon: *Discontinuities in Waveguides*, (Gordon & Breach, New York 1968)

4.16   R.E. Collin: In *Field Theory of Guided Waves* (McGraw, New York 1966)

4.17   N. Marcuvitz: *Waveguide Handbook*, MIT Rad. Lab. Ser., Vol. 10 (McGraw, New York 1951)

4.18   C.G. Montgomery, D.H. Dicke, E.M. Purcell: *Principles of Microwave Circuits*, MIT Rad. Lab. Ser., Vol. 8 (McGraw, New York 1948)

4.19   G.L. Ragan: *Microwave Transmission Circuits*, MIT Rad. Lab. Ser., Vol. 9 (McGraw-Hill, New York 1948)

4.20   S. Sridhar, D. Reagor, G. Grüner: Rev. Sci. Instr. **56**, 1956 (1985)

4.21   C.H. Papas: J. Appl. Phys. **25**, 1552 (1954)

4.22   R. Morse, H. Bohm: Phys. Rev. **108**, 1094 (1957)

4.23   J. Bowman, T. Senior, P. Uslenghi: In *Electromagnetic and Acoustic Scattering by Simple Shapes* (Hemisphere, New York 1987)

4.24   J.A. Osborn: Phys. Rev. **67**, 351 (1945)

4.25   L. Landau, E. Lifschitz: *Electrodynamics of Continuous Media* (Pergamon, Oxford 1989)

4.26   T.W. Kim, W.P. Beyermann, D. Reagor, G. Grüner: Rev. Sci. Instr. **59**, 1219 (1988)

4.27   M. Dressel, O. Klein, S. Donovan, G. Grüer: Int'l J. Infrared Millimeter Waves **14**, 2489 (1993)

4.28   K. Bechgaard, C.S. Jacobson, K. Mortensen, H.J. Pedersen, N. Thorup: Solid State Commun. **33**, 1119 (1980)

4.29   K. Bender, K. Dietz, H. Endres, H.W. Helberg, I. Hennig, H.J. Keller, H.W. Schaffer, D. Schweitzer: Mol. Cryst. Liq. Cryst. **107**, 45 (1993)

4.30   J.M. Lawrence, P.S. Riseborough, R.D. Parks: Rep. Prog. Phys. **44**, 1 (1981)

4.31   G. Stewart: Rev. Mod. Phys. **56**, 755 (1984)

4.32   H.R. Ott, Z. Fisk: In *Handbook on the Physics and Chemistry of the Actinides*, ed. by A.J. Freeman, G.H. Lander (Elsevier, Amsterdam 1987) pp. 85-225

4.33   G. Grüner: Adv. Phys. **23**, 941 (1974)

4.34   G. Grüner, A. Zawadowski: Rep. Prog. Phys. **37**, 1947 (1974)

4.35   P. Fulde, J. Keller, G. Zuricknagl: Solid State Phys. **41**, 1 (1988)

4.36   A.J. Millis, P.A. Lee: Phys. Rev. B **35**, 3394 (1987)

4.37   A.J. Millis, M. Lavagna, P.A. Lee: Phys. Rev. B **36**, 864 (1987)

4.38   P. Coleman: Phys. Rev. Lett. **59**, 1026 (1987)

4.39   A.M. Awasthi, L. Degiorgi, G. Grüner, Y. Dalichaouch, M.B. Maple: Phys. Rev. B **48**, 10692 (1993)

4.40   B.C. Webb, A.J. Sievers, T. Mihalisin: Phys. Rev. Lett. **57**, 1951 (1986)

4.41   F. Marabelli, G. Travaglini, P. Wachter, J.J.M. Franse: Solid State Commun. **59**, 381 (1986)

4.42   P.E. Sulevski, A.J. Sievers, M.B. Maple, M.S. Torikachvili, J.L. Smith, Z. Fisk: Phys. Rev. B **38**, 5338, (1988)

4.43   M. Tinkham: *Introduction to Superconductivity* (McGraw-Hill, New York 1975)

4.44   D.C. Mattis, J. Bardeen: Phys. Rev. **111**, 412 (1958)

4.45   O. Klein: Conductivity coherence factors in superconductors. Ph.D. Thesis, University of California, Los Angeles (1993)

4.46   O. Klein, E.J. Nicol, K. Holczer, G. Grüner: Phys. Rev. B **50**, 6307 (1994)

4.47   J.R. Schrieffer: *Theory of Superconductivity* (Addison-Wesley, Reading 1964)

4.48   G. Grüner: Rev. Mod. Phys. **60**, 1129 (1988)

4.49   D. Reagor, M. Maki, G. Grüner: Phys. Rev. B **32**, 8445 (1985)

4.50   A. Philipp, W. Mayr, T.W. Kim, G. Grüner: Solid State Commun. **62**, 521 (1987)

4.51   S. Donovan, Y. Kim, L. Degiorgi, M. Dressel, G. Grüner, W. Wonneberger: Phys. Rev. B **49**, 3363 (1994)

4.52   G. Grüner: Rev. Mod. Phys. **66**, 1 (1994)

# 5. Far-Infrared Fourier Transform Spectroscopy

Ludwig Genzel
With 38 Figures

This chapter is concerned with the Fourier Transform Spectroscopy (FTS) [5.1–12], especially for the Far-Infrared Region (FIR) (10–500 cm$^{-1}$) and its application to solid-state physics. FTS in its present form was introduced 1951 by *Fellgett* [5.13] who realized the potential of this spectroscopic method due to the socalled multiplex advantage. Already in 1911 *Rubens* and *Wood* [5.14] used a preliminary form of FTS including its phase information for acquiring spectra in the very far infrared. But they soon gave it up because of the computational difficulties for performing the Fourier Transform (FT). Michelson should also be named in the context of the history of FTS [5.15]. From 1891–1892 he successfully used his famous interometer to investigate the fine structure of atomic spectral lines by means of his "visibility technique", which however, did not involve the phase information and was therefore restricted to very simple line spectra.

The existence of today's fast digital computers with their large memory capacitance made the breakthrough of FTS possible during the last 25 to 30 years), as one of the most important spectroscopic methods for the InfraRed (IR) as well as for the nuclear magnetic resonance spectroscopy with pulse excitation. The IR-FTS has also not been pushed aside by laser spectroscopic techniques, as was sometimes expected. Its fast quantitative analysis of a broad spectral range, the application as a "fingerprint" method for the characteristic molecular vibrations, and also its success in solid state physics has caused the enormous extension. Since the present article is thought as only a general introduction into FTS for understanding its possible applications in solid state physics, for a more detailed study we refer to the book literature [5.1–12] and to survey articles [5.16–28].

Each experimental setup for the separation of electromagnetic radiation in its spectral components can lastly be considered as a realization of the Fourier transformation for finding the spectrum. Examples are the prism and grating-monochromators, or the Fabry-Pérot interferometry. FTS in the present sense means the measurement of the so-called interferogram $P(\gamma)$ of a two-beam interferometer as a function of the path difference $\gamma$ (phase) between the two arms of the interferometer and the computation of the spectrum $p(\nu)$ from this by Fourier transformation. It is useful here to give the frequency $\nu$ in wavenumbers (cm$^{-1}$). The interferogram is the auto- or cross-correlation of the radiation fields in the two interferometer arms under unambiguous use

Topics in Applied Physics, Vol. 74
**Millimeter and Submillimeter Wave Spectroscopy of Solids** Ed.: G. Grüner
© Springer-Verlag Berlin Heidelberg 1998

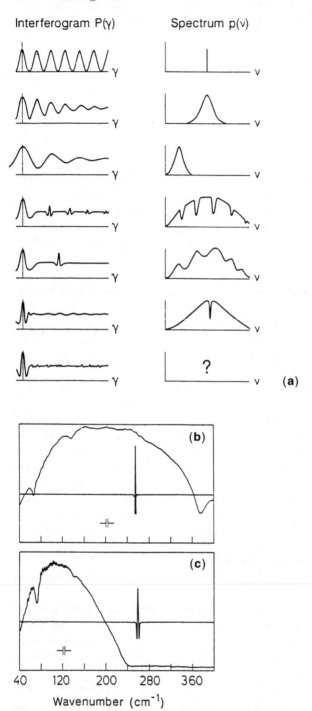

**Fig. 5.1 a–c.** Simple spectra and their interferograms

of the phase (path difference) which demands, therefore, the application of two-beam interferometers and not multiple-beam devices. The indirect way of obtaining the spectral information was the reason why in its beginnings, FTS was considered with sceptics. But meanwhile, one has realized that the advantages of FTS over slit spectroscopy are so large that the latter one is used only for special purposes. The main advantages are the already mentioned multiplex advantage and the so-called throughput advantage [5.18], which also exists for the Fabry-Pérot interferometry.

Since the interferogram function is built up by many spectral elements having different wavelengths and intensities, it is a rather complicated function of $\gamma$. Most of the commercially available FT-spectrometers, therefore, do not bother the user by looking on the interferogram. For the critical spectroscopist, however, this would be the wrong attitude, since learning "reading" the features of $P(\gamma)$ already given some insight concerning the expected spectrum. It is, therefore, useful to look first at some simple spectra $p(\nu)$ and their interferograms $P(\gamma)$ (Fig. 5.1a). A monochromatic line at $\nu_0$ in $p(\nu)$ results in a $\cos^2$ form of $P(\gamma)$. If $\nu_0$ is low, then the "wavelength" in $P(\gamma)$ is large and vice versa. If the line is Lorentzian-broadened then the $\cos^2$ function in $P(\gamma)$ is superimposed by an exponential decay function. A very broad and unstructured spectrum yields a $P(\gamma)$ which has a first minimum close to $\gamma = 0$ quickly resulting then into a constant $P(\gamma)$ (Fig. 5.1b). Thus the $P(\gamma)$ images the coherence length of the radiation field. A channeling on a broad spectrum gives a rather sharp structure in $P(\gamma)$. A sharp structure superimposed on a broad spectrum yields, on the other hand, a channeling of $P(\gamma)$. How far in $\gamma$ shall $P(\gamma)$ be measured? There is no clear answer to this question. There might still come structure in $P(\gamma)$ for larger $\gamma$-values (Fig. 5.1c). It is recommendable to make a preliminary fast scan for a rough test. The question belongs to the problem of the finite spectral resolution. Also improtant is the question of the symmetry of $P(\gamma)$ around $\gamma = 0$. How can the position of $\gamma = 0$ be determined and what accuracy is needed to do this? This belongs to the problem of the so-called phase errors. The measurement of $P(\gamma)$ is, of course, not free of noise. How is this noise transformed to the spectral noise? The scan of $P(\gamma)$ can be done continuously and slow (slow scan), in steps (step scan), or fast and repeatedly (rapid scan). The last mode is mostly used in commercial instruments. It does not need the chopping of that radiation beam and gives, therefore, more time to use it. Furthermore, rapid scanning has the advantage of modulating the different wavelengths of the spectrum, each by a different audio frequency, which is very helpful for the necessary prefiltering of $p(\nu)$.

Since the scope of this article is solid state spectroscopy, one can ask where the solid should be placed in the spectrometer. Normally, one puts the sample outside of the interferometer and well behind it, since the interferometer with its beam spitter alraydy reduces the range of $p(\nu)$ optically. In this way the sample is more decoupled thermally from the hot radiation source. But nevertheless, one should keep in mind that the sample is illuminated all the time

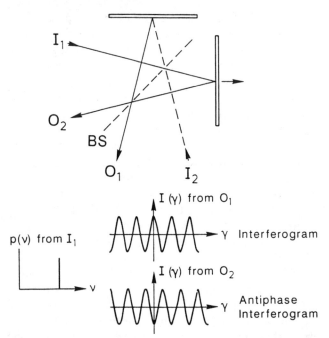

**Fig. 5.2.** The two input ports $I_1$ and $I_2$ and the two output ports $O_1$ and $O_2$ of a two-beam interferometer. Normal and antiphase interferogram

by the whole spectral range of $p(\nu)$. This is the reason why the FTS cannot be used if the sample would, show for instance, persistent photoconductivity. Sometimes the sample is placed in one arm of the interferometer (dispersive FTS [5.29]) where now the amplitude reduction and the phase influence of the sample are both investigated. One can determine the complex dielectric function in this way.

Each two-beam interferometer has two input ports and two output ports as schematically shown in Fig. 5.2. Input 1 gives a "normal interferogram" for the output 1, however, there also is an "antiphase interferogram" for the output 2, the radiation of which is normally going back to the source. Special types of interferometers allow the use of both inputs and outputs (Sect. 5.2) with a common detector for both outputs. In the ideal case, then cancel all structures in the interferogram function (optical null), as shown in Fig 5.3. If we place two nearly identical samples in the two outputs then it results in a small interferogram which contains only the spectral information of their difference (dual beam FTS).

A promising combination is that of a FT spectrometer with FIR ellipsometry [5.30, 31], thus yielding directly the complex conductivity or dielectric function of the sample. One now has, however, an oblique incidence of the radiation beam on the sample which needs the foreknowledge of a possible

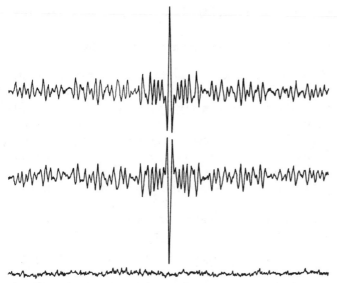

**Fig. 5.3.** The "optical null" of interferogram and antiphase interferogram

crystalline anisotropy for the right positioning of the variable polarizer. The oblique incidence also needs a larger sample size than for normal incidence.

A well-known problem for FIR spectroscopy arises in the case of highly reflecting metals. This is especially so for the high $T_c$ superconductors. Their reflectance at low temperatures might be 99% and more throughout the FIR. One then needs to know the reflectance of the necessary reference mirror of much better than 1%, which is almost impossible. Here, multiple-reflection methods in combination with FTS become essential. One such technique is described later.

Can we extend the FTS into the mm-region or even further? A limitation seems to exist here. The spectral intensity of thermal sources dies out and therefore also the attainable resolution. Synchrotron radiation sources might help to come somewhat further to lower frequencies. If we look on the range from 10 to 1 cm$^{-1}$ and if we reach a resolution there of only 1 cm$^{-1}$, one has just 10 relevant spectral points. The multiplex advantage (see later) is then only 3 in the S/N ratio. This then, is the point where coherent sources or the time domain spectroscopy has to set in.

In the following sections we will treat the most important problems discussed up to now in a more quantitative manner.

## 5.1  General Aspects of Fourier Transform Spectroscopy

### 5.1.1  Basic Theory of Fourier Transform Spectroscopy

Consider the scheme of Fig. 5.4 for a two-beam interferometer. The source S, beam splitter BS, detector D, filters, and a possible sample outside the interferometer determine the spectral intensity $p(\nu)$. For the sake of generality we also allow the sample to be in one arm (arm 2) of the interferometer instead of being outside. We call the power in arm 1 then $p_1(\nu)$, while in arm 2 the sample changes $P_1$ to a lower value $p_2(\nu)$ due to absorption. The sample will furthermore introduce a phase shift $-\phi(\nu)$ caused by its dispersive properties. In order to perform the cross-correlation between the fields in arm 1 and arm 2, we produce a path difference $\gamma$ in arm 1 (by a movable mirror) causing a phase change of $2\pi\nu\gamma$ at the frequency $\nu$. The amplitudes $a_1(\nu)$ and $a_2(\nu)$ in the two arms can then be expressed by

$$a_1(\nu) = \sqrt{p_1(\nu)}e^{i2\pi\nu\gamma}, \quad a_2(\nu) = \sqrt{p_2(\nu)}e^{-i\phi(\nu)} . \tag{5.1}$$

In (5.1) we have omitted the general and common phase contribution from the time and space dependence of the fields. We also left out the influence of the beam splitter (BS) on $a_1$ and $a_2$ because it is a common factor after passing BS the second time. The detector then measures the frequency integral of the absolute square amplitudes:

$$P(\gamma) = \int_0^\infty |a_1 + a_2|^2 d\nu = \int_0^\infty [p_1(\nu) + p_2(\nu)]\, d\nu + \int_0^\infty (a_1 a_2^* + a_1^* a_2)d\nu. \tag{5.2}$$

The first term on the right side of (5.2) is independent of $\gamma$ and $\phi$ and shall be called $P(\infty)$, since for large enough $\gamma$, the finite coherence length of the radiation let the second term in (5.2) disappear. The proper interferogram function is then defined by

$$I(\gamma) = P(\gamma) - P(\infty) = 2 \int_0^\infty \sqrt{p_1(\nu)p_2(\nu)} \cos\left[2\pi\nu\gamma + \phi(\nu)\right] d\nu. \tag{5.3}$$

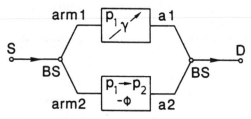

**Fig. 5.4.** Scheme of a two-beam interferometer with a dispersive sample in one arm $(p_1 \rightarrow p_2, -\phi)$ and the production of the path difference $(\gamma)$ in the other arm. S = source, D = detector, BS = beam splitter

We will now formally extend the function under the integral to negative frequencies

$$
I(\gamma) = \int_{-\infty}^{+\infty} \sqrt{p_1 p_2}\, e^{i\phi} e^{i2\pi\nu\gamma}\, d\nu
$$

$$
= \int_{-\infty}^{+\infty} (\sqrt{p_1 p_2}\cos\phi + i\sqrt{p_1 p_2}\sin\phi) e^{i2\pi\nu\gamma}\, d\nu. \tag{5.4}
$$

With the definitions of

$$
p_s(\nu) = \sqrt{p_1 p_2}\cos\phi, \quad p_a(\nu) = \sqrt{p_1 p_2}\sin\phi, \tag{5.5}
$$

and of the complex spectral power

$$
\hat{p}(\nu) = p_s(\nu) + i p_a(\nu), \tag{5.6}
$$

one has from (5.4)

$$
I(\gamma) = \int_{-\infty}^{+\infty} \hat{p}(\nu) e^{i2\pi\nu\gamma}\, d\nu. \tag{5.7}
$$

Since $I(\gamma)$ has to be trivially a real quantity, $\hat{p}(\nu)$ must fuffill the reality condition

$$
\hat{p}^*(-\nu) = \hat{p}(\nu). \tag{5.8}
$$

With (5.7) we have found the proper form for a Fourier integral transformation which yields directly $\hat{p}(\nu)$ with the back transformation

$$
\hat{p}(\nu) = \int_{-\infty}^{+\infty} I(\gamma) e^{-i2\pi\nu\gamma}\, d\gamma. \tag{5.9}
$$

If we know $I(\gamma)$ for all $\gamma$ from $-\infty$ to $+\infty$, then the complex spectrum $\hat{p}(\nu)$ can be determined by a sine and a cosine transform:

$$
|\hat{p}(\nu)| = \sqrt{p_s^2 + p_a^2} = \sqrt{p_1 p_2}, \quad \phi(\nu) = \arctan\left(\frac{p_a}{p_s}\right). \tag{5.10}
$$

For $\phi \neq 0$, it follow that $I(\gamma)$ is not symmnetrical in $\gamma$ as is already seen in (5.3). Thus, one has to measure $I(\gamma)$ two-sided. If, however, there is no dispersive medium inside the interferometer, which also includes the case that the sample is outside, then we have with $\phi = 0$ and $p_1 = p_2 = p$ a symmetrical interferogram, thus

$$
p(\nu) = \int_{-\infty}^{+\infty} I(\gamma)\cos(2\pi\nu\gamma)\, d\gamma = 2\int_{0}^{\infty} I(\gamma)\cos(2\pi\nu\gamma)\, d\gamma. \tag{5.11}
$$

Now one only needs to measure $I(\gamma)$ one-sided to get the spectrum $p(\nu)$ after the FT. Equation (5.3) shows for this case that at the position $\gamma = 0$ (the white light position) holds

$$I(0) = 2 \int_0^\infty p(\nu)d\nu.$$

The same value follows from (5.2) for $P(\infty)$. (see also Fig. 5.1), therefore:

$$I(0) = P(\infty). \tag{5.12}$$

This condition is a critical test for the right adjustment of the interferometer.

### 5.1.2 Resolution and Apodisation

With the (5.9–11) the problem seems solved, namely being able to calculate an unknown spectrum from a measured interferogram by Fourier transformation. However, $I(\gamma)$ needs to be measured for all $\gamma$ from either $-\infty$ to $+\infty$ or from 0 to $\infty$, which is, of course, impossible. One has to modify the integral relations for a finite $\gamma_{max}$. It will be shown that this yields the finite spectral resolution for all spectroscopic instruments in a somewhat modified form.

We introduce, therefore, the truncated or "observed" interferogram by

$$I_{obs}(\gamma) = I(\gamma) \cdot S(\gamma), \tag{5.13}$$

where $S(\gamma)$ is a proper symmetrical screen function which runs from $-\gamma_{max}$ to $+\gamma_{max}$. Correspondingly, we find an "observed" spectral power:

$$
\begin{aligned}
p_{obs}(\nu) &= \int_{-\infty}^{+\infty} I_{obs}(\gamma)e^{-i2\pi\nu\gamma}d\gamma \\
&= \int_{-\gamma_{max}}^{+\gamma_{max}} I(\gamma) \cdot S(\gamma)e^{-i2\pi\nu\gamma}d\gamma. 
\end{aligned}
\tag{5.14}
$$

The convolution theorem allows us to rewrite (5.14) into

$$p_{obs}(\nu) = p(\nu) * s(\nu) = \int_{-\infty}^{+\infty} s(\nu - \nu')p(\nu')d\nu', \tag{5.15}$$

with

$$s(\nu) = \int_{-\infty}^{+\infty} S(\gamma)e^{-i2\pi\nu\gamma}d\gamma = 2\int_0^{\gamma_{max}} S(\gamma)\cos(2\pi\nu\gamma)d\gamma. \tag{5.16}$$

The observed spectrum $P_{obs}(\nu)$ is equal to the true spectrum $p(\nu)$ convolved with the scanning function $s(\nu)$ which is the back FT of $S(\gamma)$. Normally

a)

$I(\gamma) \cdot S(\gamma) \longrightarrow p(v) * s(v)$

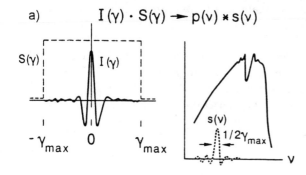

b)

$I(\gamma) \cdot S(\gamma) \longrightarrow p(v) * s(v)$

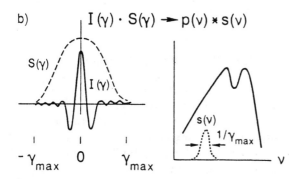

**Fig. 5.5 a,b.** Resolution and apodisation. $I(\gamma)$ = interferogram, $S(\gamma)$ = screen or apodisation function, $p(\nu)$ = true spectrum, $s(\nu)$ = scanning function

one is not going to solve the integral relation of (5.15); rather one tries to find screen functions $S(\nu)$ which reproduce the spectral features as well as possible, i.e., for instance, Lorentzians as Lorentzians.

The simplest, but really the worst case is the use of the rectangular screen (Fig. 5.5),

$$S(\gamma) = \begin{cases} 1 & \text{for} & |\gamma| \le \gamma_{\max}, \\ 0 & \text{for} & |\gamma| > \gamma_{\max}, \end{cases} \tag{5.17}$$

which yields with (5.16) the scanning function

$$s(\nu) = 2\gamma_{\max} \cdot \text{sinc}(2\pi\nu\gamma_{\max}), \tag{5.18}$$

where $\text{sinc}\,x = \sin x / x$ is the well-known diffraction function.

Equation (5.18) has a main maximum and a series of decaying side-maxima and -minima. The half width of the main maximum is

$$\Delta\nu \simeq \frac{1}{2\gamma_{\max}}. \tag{5.19}$$

The side lobes of (5.18) are rather disturbing for analyzing neighboring spectral features. As already mentioned, one has to change $S(\gamma)$ of (5.17) in the direction of a truncated Gauss function (Fig. 5.5) since the Gauss function is the only one which transforms again to a Gauss function. The procedure to reduce the side lobes of the scanning function is generally called apodisation ("putting the feet down").

The chosen screen function $S(\gamma)$ is mathematically multiplied onto the already measured and thus truncated interferogram. Commercially available FT spectrometers offer a variety of apodisation functions with their computer programs. The general result of the apodisation is a doubling of the resolution value of (5.19) (Fig. 5.5). We then have the simple result

$$\delta\nu \simeq \frac{1}{\gamma_{max}}. \tag{5.20}$$

The apodisation can also be applied to nonsymmetrical interferograms ($\phi \neq 0$) as long as $S(\gamma)$ is symmetrical to $\gamma = 0$. It should be noted that $\delta\nu$ in the case of symmetrical interferograms is independent of whether one measures them one-sided or two-sided.

### 5.1.3 Digital Analysis. Sampling: The Aliasing Problem

In today's FTS one takes the information from the interferogram only at equally spaced sampling points with a certain spacing $\Delta\gamma$. In order to define this $\Delta\gamma$, one has to have pre-information about the extension of the spectrum, i.e., the value of $\nu_{max}$. Of importance is the so-called sampling theorem of the information theory, which makes a statement about the full information for the spectrum if the interferogram is known on a discrete number of points. The statement says: If the spectrum is limited to $\nu \leq \nu_{max}$, then the interferogram function is completely determined at $\gamma$-points with distance $\Delta\gamma = 1/2\nu_{max}$. Descriptively, this means to take the interferogram at points with a distance of $\lambda_{min}/2$ where $\lambda_{min} = \nu_{max}^{-1}$. The position $\gamma = 0$ has to be one of these points. Since the Fourier theorem is reciprocal in $\nu$ and $\gamma$, we have another statement: If the interferogram function is limited to $|\gamma| \leq \gamma_{max}$ then the spectrum is completely determined at $\nu$-points with a distance of $\Delta\nu = 1/2\gamma_{max}$.

The preknowledge of $\nu_{max}$ needs a prefiltering of the spectrum. This can be done optically, although it is not easy to do without losses of the spectral power. In principle, one can apply mathematic filtering which is, however, not often in use. The most elegant way is electronic filtering. We consider the continuous scan of $I(\gamma)$ in the rapid scan mode before the application of the sampling. The path difference $\gamma$ is driven in time $t$ with constant speed $v$ of one of the interferometer mirrors

$$\gamma = 2vt. \tag{5.21}$$

In this way each frequency $\nu$ which is contained in the spectrum is "labeled" by its own modulation audio-frequency:

$$f(\nu) = 2\nu v. \tag{5.22}$$

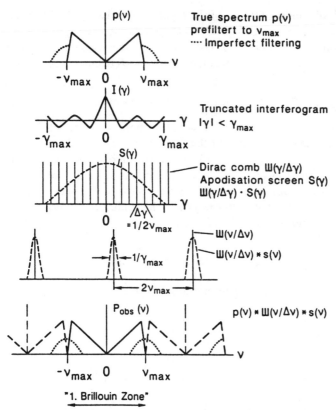

**Fig. 5.6.** Sampling of the interferogram with a Dirac comb. The Brillouin zones of the observed spectrum. *Dotted lines*: Not appropriately prefiltered according to $\Delta\gamma = 1/2\nu_{max}$

In commercial FT spectrometers $f(\nu)$ ranges from about 10 Hz to a few kHz. Therefore, a low pass electronic filtering is projected as filtering into the optical region.

If the filtering according to $\Delta\gamma = 1/2\nu_{max}$ is not properly done, i.e., if spectral power is still present above $\nu_{max}$, then one finds this radiation in $p_{obs}$ back in the range of $\nu < \nu_{max}$. One calls this the aliasing problem. We explain this and the procedure of getting $p_{obs}$ in the case of interferogram sampling schematically in Fig. 5.6. First one has the prefiltered spectrum and the resulting truncated interferogram. For the simulation of the sampling we introduce the Dirac comb function $\mathrm{III}(\gamma/\Delta\gamma)$ defined by ($n$ being integers)

$$\mathrm{III}(x) = \sum_{n=-\infty}^{+\infty} \delta(x - n). \tag{5.23}$$

the $\delta$-functions of $\mathrm{III}(\gamma/\Delta\gamma)$ shall have distances of $\Delta\gamma = 1/2\nu_{max}$ according to the sampling theorem. The FT of the Dirac comb in $\gamma$ is again a Dirac comb

$III(\nu/\Delta\nu)$ in $\nu$ with distances of $2\nu_{max}$ of the $\delta$-peaks. We also apply, however, the apodisation screen $S(\gamma)$ on $III(\gamma/\Delta\gamma)$, so the FT is then $III(\nu/\Delta\nu) * s(\nu)$, i.e., each $\delta$-function on the $\nu$-scale is converted to a scanning function of width $1/\gamma_{max}$. Finally, we take into account the truncated interferogram for $p_{obs}(\nu)$ to get:

$$p_{obs}(\nu) = FT[I(\gamma) \cdot S(\gamma) \cdot III(\gamma/\Delta\gamma)] = p(\nu) * s(\nu) * III(\nu/\Delta\nu). \quad (5.24)$$

$III(\gamma/\Delta\gamma)$ is a one-dimensional lattice of sampling points. As in solid state physics we find then one-dimensional Brillouin zones on the $\nu$-scale. The first zone goes from $-\nu_{max}$ to $+\nu_{max}$ and will be scanned by only one of the scanning peaks of $III(\nu/\Delta\nu) * s(\nu)$. In the next Brillouin zone the next peak takes over and the spectrum repeats. If we would have real spectral power above $\nu_{max}$ due to insufficient prefiltering, then this radiation is also folded back to the first zone and distorts the spectrum.

## 5.1.4 Realization of the Digital Fourier Transform

We consider now a symmetrical interferogram which has been measured one-sided between 0 and $\gamma_{max}$ with increments of $\Delta\gamma = 1/2\nu_{max}$. The number $Z$ of sampling points is therefore

$$Z = \frac{\gamma_{max}}{\Delta\gamma} = 2\nu_{max}\gamma_{max}. \quad (5.25)$$

We then calculate the same number of spectral points for $p_{obs}$:

$$
\begin{aligned}
p_{obs}(m\Delta\nu) &= 2\sum_{n=0}^{Z} I_{obs}(n\Delta\gamma)\cos(2\pi m\Delta\nu n\Delta\gamma) \\
&= 2\sum_{n=0}^{Z} I_{obs}(n\Delta\gamma)\cos(\pi\frac{m\cdot n}{Z}).
\end{aligned}
\quad (5.26)
$$

This shows that $Z^2$ computational steps have to be done. If the spectrum of interest ranges up to $500\,cm^{-1}$ with a resolution of $0.1\,cm^{-1}$ we get $Z = 10^4$, which means $10^8$ computational steps for a straightforward calculation. This is possible but unnecessary since one can apply the so-called Cooley-Tukey fast Fourier transform algorithm [5.32, 33] which will not be explained here in detail. The reader is referred to [5.9] for this.

## 5.1.5 Errors in Fourier Transform Spectroscopy

There are various specific effects which can distort the recovered spectrum or cause a reduction in intensity and resolution.

One of these effects occurs sometimes in the slow scan mode, namely, a drift in $P(\gamma)$. This drift results, however, only in a distortion of $p_{obs}$ at very low frequencies. Another effect is the occurrence of a channeling in $p_{obs}$. In trivial cases this might be caused by an electronic pulse resulting in a sharp peak in the interferogram. This can often be removed there and replaced by neighboring interferogram points. But channeling can also come from close lying plane-parallel surfaces, for instance in the detector. They again cause a sharp peak in the interferogram and can be treated as explained above. Often the interferogram is insufficiently modulated, i.e., (5.12) is not fulfilled. This acts like a general loss of spectral power and reduces the signal-to-noise ratio. It is not easy to observe this in rapid scan instruments. Here, some readjustment of the interferometer or of the imaging on the detector is necessary. The interferometer can be asymmetric although no sample is inside the interferometer. In this case one has an insufficient overlapping of the beams on the beam splitter and this effect might even change with the path difference. Here, again, a readjustment is necessary to avoid greater spectral distortions. The effect is also difficult to observe in rapid scan instruments since the interferogram appears asymmetric there anyway due to the frequency-dependent time constant of the detector (see shirping below). One has to go to the slowest possible scan.

Phase errors are the most serious errors. The Fourier transform to $p_{obs}$ needs the knowledge of the absolute value of the path difference $\gamma$ and, therefore, especially the accurate position of $\gamma = 0$. If the first sampling point fails to hit this position, then very severe distortions of the spectrum up to "negative intensities" can occur. If $\gamma = 0$ is not a sampling point, then an otherwise symmetrical interferogram appears asymmetric. We have shown that asymmetric interferograms can be treated by a cosine- and a sine-transformation. This is done therefore, in all modern FT spectrometers: Around a small portion of the approximate position $\gamma = 0$ one takes a two-sided interferogram. If the sine-transform is not zero, one corrects the $\gamma$-scale such that it disappears. This is done automatically. We already mentioned the shirping of the interferogram in the rapid scan mode due to the finite response time of the detector. This shirping is really a positive effect: The higher audio-frequencies which belong to higher optical frequencies are delayed and reduced in amplitude. In this way the grand maximum is smeared out and the dynamical range problem is reduced. The shirping distortion of the interferogram has to be corrected, however, before the digital transform. The above mentioned procedure of using the sine- and cosine-transform is now applied for each modulation frequency and separately corrected.

### 5.1.6 Noise in Fourier Transform Spectroscopy

Consider the noise amplitude $N(\gamma)$ in the interferogram. After truncation with the screen function $S(\gamma)$ we can transform it to the noise spectrum:

$$n(\nu) = \int\limits_{-\infty}^{+\infty} N(\gamma)S(\gamma)e^{-i2\pi\nu\gamma}d\gamma.$$

Here we assume that $N(\gamma)$ is independent of the whole interferogram signal on which it is superimposed. This means that the noise should not originate in the source. For large enough $\gamma_{max}$ and $\nu_{max}$ holds $\overline{N(\gamma)} \approx 0$ and $\overline{n(\nu)} \approx 0$. One therefore considers the quantity $|n(\nu)|^2$ which has the convolution product $N(\gamma)S(\gamma) * N(\gamma)S(\gamma)$ as FT.

Thus

$$\int\limits_{-\infty}^{+\infty} |n|^2 e^{-i2\pi\nu\gamma} d\nu = \int\limits_{-\infty}^{+\infty} N(\gamma')S(\gamma')N(\gamma-\gamma')S(\gamma-\gamma')d\gamma'.$$

We take the integrals best at $\gamma = 0$ since this is no special point for the noise:

$$\int\limits_{-\infty}^{+\infty} |n(\nu)|^2 d\nu = \int\limits_{-\infty}^{+\infty} |N(\gamma)S(\gamma)|^2 d\gamma, \tag{5.27}$$

which is Parseval's theorem. Since the noise should be blocked outside of $\nu_{max}$ and $|\gamma_{max}|$ we find

$$\int\limits_{-\nu_{max}}^{+\nu_{max}} |n(\nu)|^2 d\nu = \int\limits_{-\gamma_{max}}^{+\gamma_{max}} |N(\gamma)S(\gamma)|^2 d\gamma. \tag{5.28}$$

We now define average values:

$$\overline{n^2} = \frac{1}{2\nu_{max}} \int\limits_{-\nu_{max}}^{+\nu_{max}} |n(\nu)|^2 d\nu, \quad \overline{N^2} = \frac{1}{2\gamma_{max}} \int\limits_{-\gamma_{max}}^{+\gamma_{max}} |N(\gamma)S(\gamma)|^2 d\gamma, \tag{5.29}$$

thus

$$\nu_{max}\overline{n^2} = \gamma_{max}\overline{N^2}.$$

We have then

$$\sqrt{\overline{n^2}} = \sqrt{\frac{\gamma_{max}}{\nu_{max}}} \sqrt{\overline{N^2}}. \tag{5.30}$$

The rms noise in the spectrum increases with the rms noise in the interferogram but also with the square root of the maximal path difference, which determines the resolution in the spectrum. This is in contrast to slit monochromators where the spectral S/N ratio decreases with the square of the resolution.

It should be mentioned that the treatment of the noise transform used here is not more appropriate if the spectrum $p(\nu)$ is composed of sharp lines.

### 5.1.7 Advantages of Fourier Transform Spectroscopy

As already mentioned in the introduction, the main advantages are the multiplex advantage [5.13], and the throughput advantage [5.18] over slit monochromators. The throughput advantage holds also for the Fabry-Pérot interferometry due to the axial symmetry of its optical arrangement. Further advantages of the FTS concern the spectral filtering and the spectral noise versus resolution, as shown previously.

The multiplex advantage results from the property of FTS that all spectral elements of the spectrum $p(\nu)$ under investigation contribute all the time of the measurement to the interferogram during its scan, while slit spectrometers as well as the Fabry-Pérot interferometry scan the spectral elements one after the other. Assume we have $Z$ spectral elements, centered at $\nu_1$ to $\nu_z$ (Fig. 5.7) and having the same width and intensity. In the slit instrument we might need a time $\tau$ to measure the element with a given signal-to-noise ratio (S/N). We then need the time $T = Z \cdot \tau$ to measure the whole spectrum. An FT spectrometer having the same light throughput (below) and detector

**(a)**

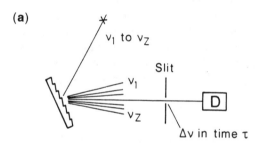

Slit-Spectrometer

**(b)**

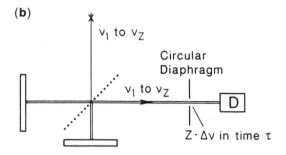

Fourier-Interferometer

**Fig. 5.7 a,b.** The multiplex advantage of FTS compared to slit spectrometers

sensitivity already yields the same S/N for all $Z$ elements in only the time $\tau$. If we allow, however, the time $T = Z \cdot \tau$ we can repeat the measurement $Z$ times, thus getting a statistical advantage of $\sqrt{Z}$ (more precisely $\sqrt{Z-1}$). If the spectrum extends up to $500\,\mathrm{cm}^{-1}$ with a width of $0.1\,\mathrm{cm}^{-1}$ for each spectral element, the advantage would be roughly 70. It should be noted, however, that the advantage exists only if the noise is independent of the underlying intensity of the function $P(\gamma)$ in the interferogram, and especially there in its grand maximum at path difference zero. In visible spectroscopy one usually uses photon counting detection systems where the statistical noise increases with the square root of the radiation power. In this case, the multiplex advantage breaks down while in IR-FTS it exists since the noise usually originates in the detector.

The following discussed throughput advantage over slit spectrometers exists for all spectral instruments with axial beam symmetry. They allow, therefore, for a circular diaphragm at the source and the detector which are imaged to each other. The central beam of the FT spectrometer suffers a phase difference at frequency $\nu$ between the two interferometer arms of $2\pi\nu\gamma$, while a beam from the edge of the source diaphragm goes through the interferometer with a certain angle $\alpha$ against the axis and causes a phase difference $2\pi\nu\gamma \cos\alpha \simeq 2\pi\nu\gamma(1 - \alpha^2/2)$ in the small angle approximation. If the central beam causes a maximum due to constructive interference at the detector we find a first ring with destructive interference if $\pi\nu\gamma\alpha^2$ equals $\pi$, which means $\alpha^2 = (\nu\gamma)^{-1}$. For a spectrum from 0 to $\nu_{\max}$ this happens first for the highest frequency $\nu_{\max}$ and for the maximal path difference $\gamma_{\max}$. We have then with $\Delta\nu = 1/2\gamma_{\max}$ that $\alpha_{\min}^2 = 2\Delta\nu/\nu_{\max}$ and a corresponding solid angle of

$$\Omega_{\min} = \pi\alpha_{\min}^2 = 2\pi\frac{\Delta\nu}{\nu_{\max}}. \tag{5.31}$$

For a more detailed derivation of (5.31) the reader is referred to [5.9].

We now consider the light power throughput or étendue $E$ through the spectrometer. If there is no absorption or scattering then $E$ is conserved. At the entrance we have (Fig. 5.8) for instance,

$$E = A\Omega_{\mathrm{s}}, \tag{5.32}$$

where $A$ is the area of the collimating mirror (or lens) of focal length $f$, and where

$$\Omega_{\mathrm{s}} = \frac{A_{\mathrm{s}}}{f^2}, \tag{5.33}$$

with $A_{\mathrm{s}}$ being the area of the source diaphragm. For the FT spectrometer we have to choose $A_{\mathrm{s}}$ so that $\Omega_{\mathrm{s}}$ equals $\Omega_{\min}$ of (5.31). This gives the étendue

$$E_{\mathrm{I}} = 2\pi\frac{\Delta\nu}{\nu_{\max}} \cdot A. \tag{5.34}$$

For a slit spectrometer is $\Omega_{\mathrm{s}} = \ell \cdot s/f^2$ where $\ell$ and $s$ are the slit-length and -width, respectively. For a smooth continuous spectrum with a nearly

(a)      Slit-Spectrometer

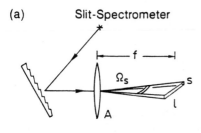

$$E_s = A\Omega_s = \frac{l}{f}\frac{\Delta v}{v_{max}}A$$

(b)      Fourier-Interferometer

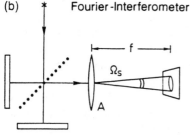

$$E_I = A\Omega_s = 2\pi\frac{\Delta v}{v_{max}}A$$

**Fig. 5.8 a,b.** The throughput (étendue) advantage of a FT interferometer compared to a slit spectrometer

constant spectral dispersion of the spectrometer $s/f \approx \Delta\nu/\nu_{\mathrm{max}}$ [5.9]. With (5.32) we now find

$$E_\mathrm{s} \simeq \frac{\ell}{f}\frac{\Delta\nu}{\nu_{\mathrm{max}}}A \tag{5.35}$$

for the same $A$ as in (5.34). Having a FIR grating spectrometer with $\ell = 1\,\mathrm{cm}$, and $f = 30\,\mathrm{cm}^{-1}$, and the same $\Delta\nu/\nu_{\mathrm{max}}$ as for an FT spectrometer, one finds $E_\mathrm{I}/E_\mathrm{s} \approx 180$.

## 5.2 Interferometers for Fourier Transform Spectroscopy

Quite a few versions of two-beam interferometers have been designed for FTS [5.4]. Among them is the Lamellar grating interferometer (Fig. 5.9, [5.1, 9]) which played a larger role in the exploration period of FTS. It can still be advantageous at the border from FIR and mm-waves because of its good beam splitting efficiency. Above $\approx 100\,\mathrm{cm}^{-1}$, however, precision difficulties set in, and below $5\,\mathrm{cm}^{-1}$ a wave guide cutoff effect between the two subgratings suddenly reduce the beam splitting effect. Lastly, it does not have the throughput advantage. For the high-resolution gas FTS one uses interferometers of the Michelson-type (Fig. 5.10a), but with path-differences $\gamma$ between

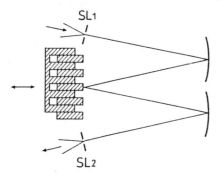

**Fig. 5.9.** The lamellar grating two-beam interferometer. SL1 = entrance slit, SL2 = exit slit in the position of the spectrum of 0$^{th}$ order. In order to avoid shadowing, the grating is turned by 90° against the axis given by the incoming and outgoing beam

the two arms of up to several meters. The necessary precision for the mirror shift is reached today by laser guiding and piezoceramic adjustments of the mirror. In solid-state spectroscopy highest resolution generally is not so important, $\delta\nu$ of 0.1 to 0.05 cm$^{-1}$ is mostly sufficient, while the photometric accuracy plays a larger role. Since the thermal sources have a rather weak spectral intensity in the FIR, the mercury arc and the SiC rod (globar) are widely used, and one is interested in a large étendue and in efficient beam dividers. For the latter, free-suspended dielectric films from materials with low absorption and high-refractive indices (for instance, Mylar) are in use. With an appropriate thickness they act as dividers due to the widely spaced multiple beam interference pattern [5.9]. The at least five octaves spanned by the FIR need an easy exchange of such sheets with different thickness. The angle of incidence should be low to avoid polarization effects and a focus on the BS would reduce its size drastically. Also metal mesh and metal grid BS are in use. The latter have very good properties for polarizing interferometers due to their almost perfect efficiency in a rather wide spectral range.

In the following we describe a few interferometers with their advantages and disadvantages and their use in special types of applications (Fig. 5.10).

a) The Michelson interferometer:
It is most widely used in commercial instruments. Its advantage is the compactness of the interferometer itself with short interferometer arms. Its disadvantage is the rather large size of the BS, and, moreover, the polarization effects occurring on it due to a beam incidence of 45°.

b) The Bruker IFS-113v interferometer (Genzel interferometer):
It provides a focus on the BS which, therefore, can be small and allows for an easy exchange under vacuum. The small beam incidence almost avoids the mentioned polarization effects. A double mirror (DM) is shifted in the rapid scan mode within the collimated beam of the interferometer. A shift $x$ of the DM causes a path difference $\gamma$ of $4x$, which is advantageous for the long wavelengths in the FIR. The interferometer is not adaptable to the highest resolution. Some disadvantage comes from the rather long interferometer arms which, therefore, have to be held under the same and constant temperature during a multiple rapid scan run.

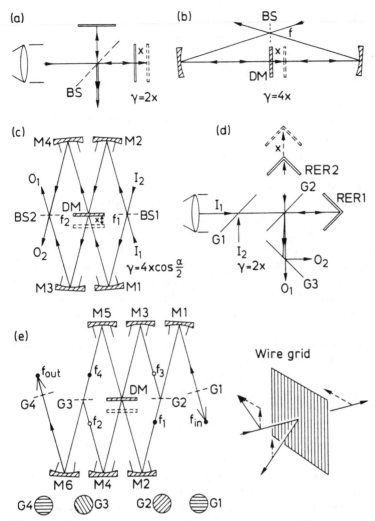

**Fig. 5.10 a-e.** Several two-beam interferometers for FTS, explanation in text

The next described interferometers are for special purposes, and therefore were not used until now in commercial instruments.

c) This type of interferometer allows the use of both input ports $I_1$ and $I_2$, and of both output ports $O_1$ and $O_2$. Thus, it can be used for the dual beam FTS. $O_1$ is the normal interferogram output for $I_1$ and the antiphase interferogram output for $I_2$, and vice versa for $O_2$, see also Fig. 5.2. Using all inputs and outputs, quite interesting compensating effects can already be achieved for the interferogram with the use of only one common detector [5.34, 35]. If one replaces BS2 on focus $f_2$ by two small plane mirrors which send the two beams back into the arms of the interferometer, and if $I_1$ is the input and $I_2$ is the output, one finds that almost all mirrors inside the interferometer are tilt compensated [5.35]. If one of the two reversing mirrors at $f_2$ is replaced by a reflection sample, the instrument could be used for dispersive FTS, see Sect. 5.3.3. A small disadvantage of the interferometer is the perpendicular shift of the beams on the two mirrors M3 and M4 if the double mirror DM is moved. For M1–M4 one can use spherical mirrors since their arrangement with the DM is Czerny-Turner coma-compensated for the coma distortion.

d) This instrument of one of the several versions of the Martin-Puplett polarizing interferometer [5.36, 37]. It is built up by 3 wire grids G1, G2, and G3, and by two roof-edge reflectors RER1 and RER2. G1 polarizes the input beam under 45° against the plane of drawing. G2 then splits this beam under proper alignment in two perpendicularly polarized beams towards RER1 and RER2, both of which have their roof edges turned at 45° against the plane of drawing. The beams then come back to G2, but each turned with its polarization by 90%. Thus the beam, which originally was reflected on G2, is now transmitted and vice versa. Grid G3 acts as an analyzer for the recombined beams which are elliptically polarized in dependence of the shift of RER2. This yields an interferogram as in a normal interferometer. The interferometer also has, of course, two input and output ports. The grids used as polarizers, beam splitter, and analyzers are almost perfect in a wide spectral range, namely, as long as the wavelength of the radiation exceeds about 5 times or more the grid spacing. The Martin-Puplett interferometer was, and is, the most successful instrument for the astrophysical investigation of the cosmic background radiation.

e) This instrument is also a polarizing interferometer [5.38] but is now designed for solid state spectroscopy, especially in the dispersive transmittance mode. It has some similarity with the interferometer of Fig. 5.10c, but it provides instead 4 focal points, $f_1$ to $f_4$, inside the interferometer due to the imaging properties of the spherical mirrors $M_1$ and $M_6$. Furthermore, it uses two grid polarizers G1 and G4, which have their wires parallel to the plane of drawing. The two grids G2 and G3 mark the beginning and the end of the interferometer itself. They are turned under 45° and 135°, respectively, against the directions of G1 and G4. In this way, again one finds an elliptically polarized beam towards G4 depending on the shift at the double mirror DM. After G4 one then finds the interferogram as in the Martin-Puplett interferometer. The focal points $f_1$ and/or $f_4$ are appropriate for samples, and eventually have cryostats to apply the dispersive FTS. All mirrors M1–M6 can be spherical since they are placed for a doubly Czerny-Turner coma-compensated instrument. If the instrument would be used in the mm-range with coherent sources [5.38], then it is necessary to under-illuminate the mirrors M2 to M5 in order to avoid diffraction effects at the mirror edges which are frequency- and path difference dependent.

## 5.3 Special Topics for Solid State Applications of Far Infrared Region-Fourier Transform Spectroscopy

This section deals with several selected examples of solid state spectroscopy where the FIR-FTS was, and is, of importance. Since the topics have been taken from the region of interest of the author, there is no claim of any survey.

### 5.3.1 Photothermal Ionization Spectroscopy

Photothermal Ionization Spectroscopy (PTIS) is a very sensitive method for detecting shallow impurities in semiconductors such as Si or Ge [5.39–42]. Figure 5.11 gives the scheme for a donor impurity with hydrogen-like electronic states in the gap below the conduction band. At low temperatures only the ground state is occupied. A photon could excite the electron from there to the conduction band causing the onset of photoconductivity. One is, however, interested in finding the energetic positions of the excited states of the impurity because this might allow identification of it. A photon $\hbar\omega$ can lift, for instance, the electron to the first excited state. This would yield an absorption. If we have, however, a density of only $10^9$ to $10^{12}$ impurities per cm$^3$, and this is the density of interest for PTIS, then it is hopeless to detect the extreme weak absorption. Instead, by having a slightly elevated temperature ($\approx$ 10–20 K), an existing low-frequency phonon $\hbar\Omega$ might lift the electron from the first excited state to the conduction band causing a photo-conducting signal. Even for impurity concentrations as given above, this signal can now be detected because currents can be measured many orders of magnitude better than an optical absorption. If the impurities have

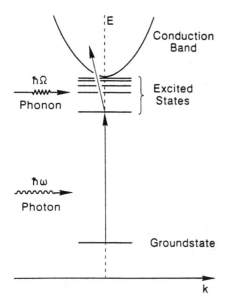

Fig. 5.11. Scheme for the photothermal ionization spectroscopy

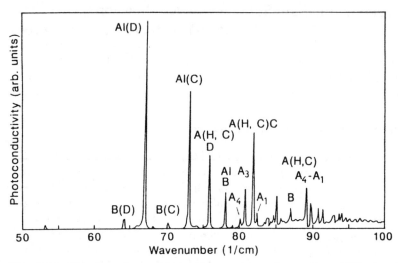

**Fig. 5.12.** The photothermal ionization spectrum of pure Ge at 10 K, showing some hydrogen-like line series of shallow impurities of Al and B

higher densities then their wavefunctions overlap and broaden the states to bands. Therefore, only very low concentrations are of interest. It has been estimated that one can just detect $10^7$ to $10^8$ impurities per $cm^3$ in extremely clean Ge, perhaps the highest sensitivity of all for finding impurities.

PTIS is very well adaptable to FTS in all cases of a shorter lifetime of the carriers in the conduction (or valence) band. The sample is now the detector and the photocurrent is Fourier transformed. Figure 5.12 shows such a measurement [5.43] on a Ge sample of very high purity (from E.E. Haller, Univ. of California at Berkeley). One finds several sharp series of lines which stem from Al- and B-impurities the concentration of which was about $10^{10}$ $cm^{-3}$. The linewidths of about $2\,cm^{-1}$ is determined solely by the resolution of the FT spectrometer. A survey of such PTIS measurements can be found in [5.41].

### 5.3.2 Amorphous Ge and Si

Pure crystalline Ge and Si do not show a first-order phonon absorption in the IR. The amorphous material, however, shows an activated phonon absorption from all branches throughout the Brillouin zone. This activation comes from local charges due to dangling bonds. Since almost all phonons can be activated, one expects spectra which represent the phonon density of states in the amorphous material.

A measurement for this has been done with the dual beam FTS using an interferometer of the type shown in Fig. 5.10c [5.44]. Two equal and plane-parallel pieces of the crystalline material, placed in the two output beams, yield an "optical null" of the interferograms as shown in Fig. 5.3. All common spectral properties, including the very strong spectral channeling due to the

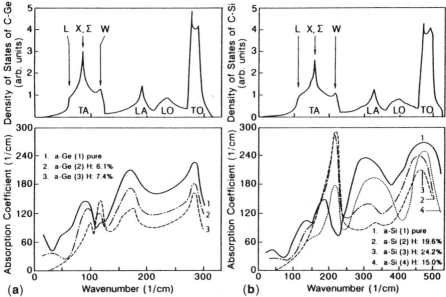

**Fig. 5.13 a,b.** The absorption spectra of amorphous Ge and Si, pure and H-doped. For comparison, the phonon density of states of crystalline Ge and Si

high refractive indices, are compensated. From one of the two pieces a layer of about 10 μm was edged off and instead a 10 μm layer of the amorphous material was sputtered on. A resulting small difference interferogram then yielded after FT the spectra shown in Fig. 5.13 [5.44]. Also shown are the known phonon density of states of the corresponding crystals. One clearly observes features in the spectra which are connected to the van Hove singularities of the crystalline density of states. H-doping, which saturates some dangling bonds, reduces the amorphous absorption. Only at about 200 cm$^{-1}$ in $a$-Si and at 117 cm$^{-1}$ in $a$-Ge, does one find an increase with H-doping, which has been interpreted as resonant band mode absorption [5.44].

### 5.3.3 Far Infrared Region Spectra of InSb

This is an example for the dispersive FTS in the reflection mode which should first be explained with the help of Fig. 5.14. We replace one of the plane mirrors of a Michelson interferometer by the reflection sample. A sharp spatial light amplitude pulse from a broad spectrum shall be split by the BS into two pulses which reach sample S and mirror M at the same time when the two arms of the interferometer are first equal in length. After reflection we find the amplitude in the arm of S reduced and also delayed, as compared to the pulse in the arm of M. The delay could be so large that no overlap of the two pulses happens after passing the BS again. Looking at the interferogram we first find the interference pattern of the mirror M

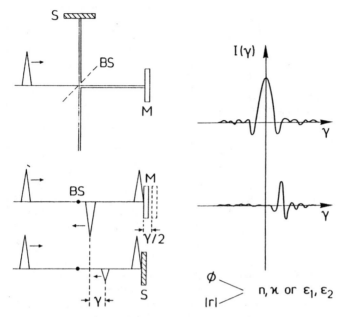

**Fig. 5.14.** Scheme for the dispersive reflection FTS. ($M$: mirror, $S$: reflection sample, $BS$: beam splitter, $I(\gamma)$: interferogram) Explanation in text

is shifted outward to produce a certain path difference $\gamma$. Since the phase delay is furthermore frequency-dependent, we find the interferogram strongly asymmetric. A real example for this is shown in Fig. 5.15. As explained in Sect. 3.1.1 with (5.10) one now applies a sine- and cosine-transformation to get the reflectance $|r|^2$ and the phase $\phi$, and from these with known formulas (Appendix 5.AI) the complex refractive index $\hat{n} = n + \mathrm{i}k$ or the complex dielectric function $\hat{\varepsilon} = \varepsilon_1 + \mathrm{i}\varepsilon_2$. The result of such a dispersive measurement on an InSb crystal is shown in Fig. 5.16 [5.45]. One observes the fundamental phonon mode (reststrahlen band) at about $175\,\mathrm{cm}^{-1}$ and,

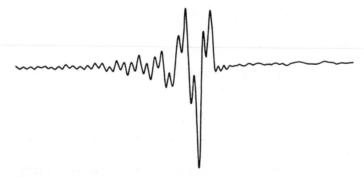

**Fig. 5.15.** The asymmetric interferogram of the dispersive FTS of InSb (to Fig. 5.16)

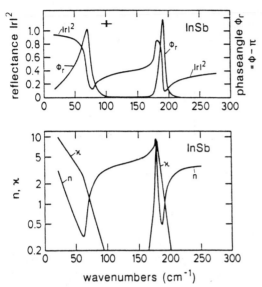

**Fig. 5.16.** The dispersive FTS of InSb. *Upper half:* Reflectance $|r|^2$ and reflection phase $\phi_r = \phi - \pi$, *lower half:* The derived spectrum of the refractive index $n(\nu)$ and of the extinction index $\kappa(\nu)$

additionally, the onset of a plasma edge below $100\,\mathrm{cm}^{-1}$ (the sample was at room temperature). It is noteworthy that the dispersive reflectance FTS is especially accurate at spectral places with low reflection. If the reflection amplitude at a level of 0.01 can be measured with an accuracy of 10% of its value, then this means it is possible to measure the reflectance $|r|^2$ of $10^{-4}$ to an accuracy of 20% of its value. If, in contrast to this, the reflection amplitude $|r|$ approaches 1 and the corresponding phase $\phi$ goes to $-\pi$, then the method fails because of inacceptable errors in the evaluation of $n$ and $k$, even for moderate errors in $|r|$ and $\phi$ (Appendix 5.AI). The dispersive reflection FTS is not applicable, therefore, for the high $T_c$ superconductors.

### 5.3.4 Far Infrared Region Spectra of SrTiO₃

SrTiO₃ belongs to a group of ferroelectric perovskites which have attracted great interest from a lattice dynamical point of view [5.46]. SrTiO₃ itself does not become ferroelectric under normal pressure, but its lowest optically active phonon mode is a typical soft mode which shifts down with temperature. SrTiO₃ is also an important substrate for films of the high $T_c$ material YBa₂Cu₃O₇ (Sect. 5.3.7). We have, therefore, remeasured it by FTS. The temperature dependence of the reflectance (Fig. 5.17) clearly shows the softening of the lowest TO mode. At $10\,\mathrm{K}$ it occurs at about $31\,\mathrm{cm}^{-1}$, yielding there a reflectance $R$ of 96–97%. The fit shows that $R$ goes down to only 92% if the frequency approaches zero. There are three other TO modes which occur for $10\,\mathrm{K}$ at $171\,\mathrm{cm}^{-1}$, $437.5\,\mathrm{cm}^{-1}$ and $553\,\mathrm{cm}^{-1}$, respectively. An almost

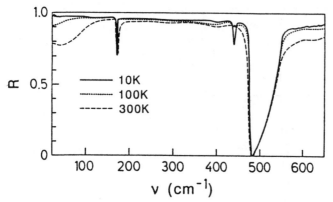

**Fig. 5.17.** FIR reflectance spectra of crystalline SrTiO$_3$, (100) cut, for three different temperatures

perfect zero of the reflectance at $482\,\mathrm{cm}^{-1}$ indicates a very large oscillator strength of the soft mode since its longitudinal optical mode, according to the Lyddane-Sachs-Teller relation, should occur there. In Fig. 5.18 the spectral dependence of the dielectric function, $\varepsilon_i$ and $\varepsilon_2$, from a fit of the $10\,\mathrm{K}$ reflectance of Fig. 5.17 is shown. The fit uses Lorentzians for the contribution of the various phonons to the complex dielectric function $\hat{\varepsilon} = \varepsilon_1 + \varepsilon_2$ or to $\hat{\sigma} = \sigma_1 + \sigma_2$ given by

$$\hat{\varepsilon} - \varepsilon_\infty = \sum_j \frac{S_j \omega_j^2}{\omega_j^2 - \omega^2 - \mathrm{i}\omega\gamma_j}, \quad \hat{\sigma} = \frac{1}{\mathrm{i}} \frac{\omega\hat{\varepsilon}}{4\pi}(cgs), \tag{5.36}$$

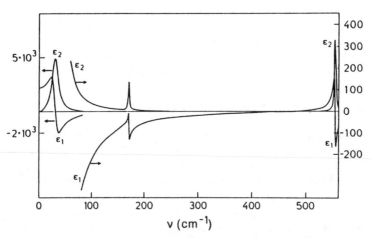

**Fig. 5.18.** Fit of the $10\,\mathrm{K}$ reflectance curve of SrTiO$_3$ of Fig. 5.17 to the dielectric functions $\varepsilon_1(\nu)$ and $\varepsilon_2(\nu)$. Note the changes of the vertical scales. The phonons given with their position $\nu_j$ [cm$^{-1}$], their oscillator strength $S_j$ and half width $\gamma_j$ [cm$^{-1}$]

where $S_j$, $\omega_j$ and $\gamma_j$ are the strength, eigenfrequency and damping, respectively, and where $\varepsilon_\infty$ is the high frequency contribution due to electronic band-band transitions.

One observes in Fig. 5.17 that the large reflectance of $SrTiO_3$ from 200 to $450\,cm^{-1}$ is caused by a moderately negative $\varepsilon_1$ together with a very small $\varepsilon_2$ above $200\,cm^{-1}$. At about $480\,cm^{-1}$ one finds the change of sign for $\varepsilon_1$, thus causing the minimum of $R$. In case of the use of $SrTiO_3$ as a substrate for high $T_c$ films, for FIR measurements one needs below $200\,cm^{-1}$ either to have the film thick enough to be opaque or to know rather exactly the dielectric function of $SrTiO_3$ and its $T$-dependence.

### 5.3.5 Phonons of Crystalline $YBa_2Cu_3O_6$

$YBa_2Cu_3O_6$ is an anti-ferromagnetic insulator. Through oxidation it can be transferred to the well-known high $T_c$ superconductor $YBa_2Cu_3O_7$ (Sect. 5.3.6). The crystal is tetragonal, i.e., the crystallographic directions $a(100)$ and $b(010)$ are equivalent, while the direction $c(001)$ is strongly anisotropic to $a$ and $b$. The material is rather ionic, thus yielding optically active phonons from the center of the Brilloum zone: One expects five such phonons for the electric vector $E\|c$ and six phonons for $E \perp c$ [5.47]. Figure 5.19 shows the reflectance measurement on a single crystal of $YBa_2Cu_3O_6$ of the size of only $2 \times 1 \times 0.6\,mm$ for $10\,K$ and $300\,K$ between $50\,cm^{-1}$ and $700\,cm^{-1}$ [5.48, 49]. One phonon for $E \perp c$ is missing. The lattice dynamical calculation expects it at $105\,cm^{-1}$ with a very low oscillator strength. The measurement is a

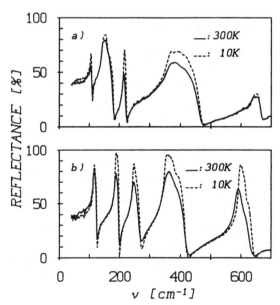

**Fig. 5.19 a,b.** FIR reflectance spectra of a $YBa_2Cu_3O_6$ single crystal at $10\,K$ and $300\,K$ and for the electric field $\|c$-axis and $\perp c$-axis

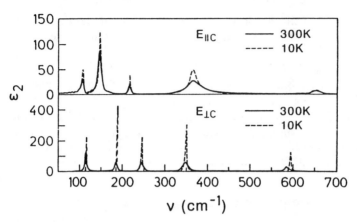

**Fig. 5.20.** Imaginary part $\varepsilon_2(\nu)$ of the dielectric function of an $YBa_2Cu_3O_6$ single crystal according to a Kramers-Kronig analysis of the reflectance curves of Fig. 5.19

good example for the application of a Kramers-Kronig transformation (Appendix 5.A). With this one first determines from $R(\nu)$ the reflection phase $\phi_r(\nu)$ and then from both the dielectric functions. Figure 5.20 shows the imaginary part $\varepsilon_2(\nu)$ of it. One observes partly a rather strong temperature dependence of the phonon half widths as well as a surprisingly strong downward shift of the highest phonon for $E \perp c$. The eigenvector of this phonon is concentrated on two oxygens of each of the two $CuO_2$ planes of the unit cell [5.46, 47]. One might expect that anharmonicity is the reason for these effects. No increase of $\varepsilon_2$ below $100\,cm^{-1}$ is observed, thus showing the insulating character of the crystal.

### 5.3.6 Spectra and Phonons of an $YBa_2Cu_3O_{7-\delta}$ Single Crystal

For this superconductor the same crystal as described in Sect. 5.3.5 has been used. It was loaded for several weeks with oxygen. Nevertheless, the oxygen deficiency $\delta$ remained at 0.2 and susceptibility measurements yielded a $T_c$ of only $86\,K$. The reflectance spectra [5.49] on this tiny crystal are shown in Fig. 5.21 for $E \perp c$ and $E||c$ and for $T = 10\,K$ and $100\,K$ in the range from 50 to $8000\,cm^{-1}$. The crystal is now orthorhombic due to the chains in the unit cell, but was twined concerning the $a$ and $b$ directions. For $E \perp c$ one finds for $T = 10\,K$ the extremely high reflectance typical for the high $T_c$ cuprates. Below about $200\,cm^{-1}$ one cannot trust any longer the measurement for the application of a Kramers-Kronig analysis. One has to apply a multi-reflection method as it is shown in Sect. 5.3.7. Therefore, we concentrate only on the $E||c$ results. A Kramers-Kronig analysis to the dynamical conductivity $\hat{\sigma} = \sigma_1 + i\sigma_2$ is shown in Fig. 5.22 [5.49]. The spectra are dominated by phonons which could be reasonably explained by lattice dynamical calculations with the exception of a very strong phonon at $152\,cm^{-1}$. We will come back to this later.

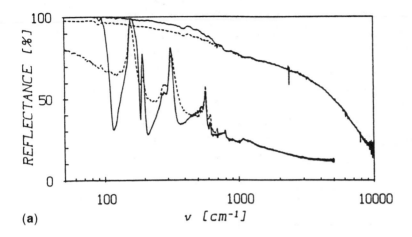

**(a)**

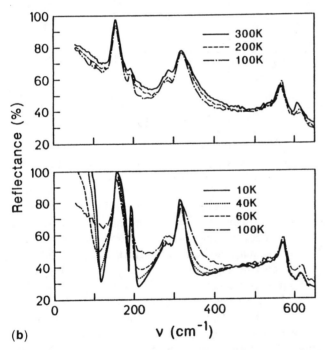

**(b)**

**Fig. 5.21 a,b.** Reflectance curves of a single crystal (twinned) $YBa_2Cu_3O_{6.8}$ at 100 K (*dashed lines*) and 10 K (*full line*) for $E \perp c$-axis and $E \| c$-axis. Explanation in text

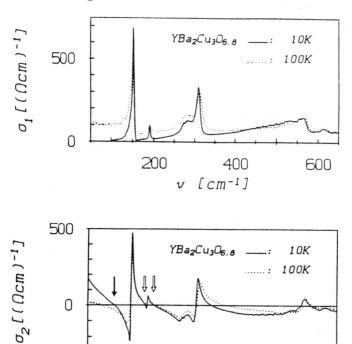

**Fig. 5.22.** Real part $\sigma_1(\nu)$ and imaginary part $\sigma_2(\nu)$ of the dynamical conductivity of the $E\|c$ spectra of Fig. 5.21 after a Kramers-Kronig analysis

Normally, one does not observe optically active phonons in metals because they are screened by the adiabatically fast movement of the free carriers. Although the crystal here is a rather bad metal with only about one free carrier (hole) per unit cell, the mentioned screening works fully for the case that the field is parallel to the $CuO_2$ planes ($E \perp c$), as seen in Fig. 5.21. For $E\|c$, however, the screening is not more effective because of the very low dynamical conductivity in this direction.

We made a rather involved fit to the Figs. 5.21, 22. First, the phonons with their position, strength, and width are taken from $\sigma_1(\nu)$. Second, we have some characteristic zeros in $\sigma_2(\nu)$, marked with arrows in Fig. 5.22. The lowest zero for 10 K is of special interest since it causes a strong minimum at about 110 cm$^{-1}$ in the reflectance. Below this minimum there occurs a rapid increase of $R$ towards 100% at 10 K, while for $T = 100$ K the $R$ curve is slowly increasing, with a slope more typical for a Drude absorption of free carriers. The minimum of $R$ at 110 cm$^{-1}$ and the corresponding zero of $\sigma_2$ at 10 K is, therefore, caused by a plasmon-phonon edge. The negative contribution to $\sigma_2$ from the phonon at 152 cm$^{-1}$ is compensated by the positive contribution

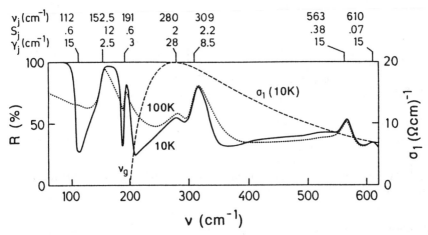

**Fig. 5.23.** A fit of the $R(\nu)$ curves ($E\|c$) of Fig. 5.21 with 7 phonons given with their position $\nu_j$ [cm$^{-1}$], oscillator strength $S_j$ and halfwidth $\gamma_j$ [cm$^{-1}$]. The very weak conductivity $\sigma_1$ for 10 K is shown starting from the gap at 200 cm$^{-1}$. Explanation in text

from the inductive supercurrent. The two other characteristic zeros of $\sigma_2$ and their corresponding minima in $R$ at 10 K are mainly due to a single gap which has to be placed at 200 cm$^{-1}$. A single gap in the $c$ direction is to be expected from calculations of the Fermi surface [5.50], because the $c$-direction cuts it in only a small portion near the $Y$-point of the reciprocal unit cell. For the fit one, furthermore, has to assume an unscreened plasma frequency of only 1000 cm$-1$, and a scattering rate of 150 cm$^{-1}$ for 10 K and of 250 cm$^{-1}$ for 100 K. Finally, the fit needs two broader electronic bands in the FIR (1st band at 450 cm$^{-1}$ with strength 1.5 and width 150 cm$^{-1}$, 2nd band at 530 cm$^{-1}$ with strength 1 and width 75 cm$^{-1}$ for 10 K), a mid-IR band (at 1800 cm$^{-1}$ with strength 12 and width 1200 cm$^{-1}$), and an $\varepsilon_\infty$ of 4.8. The result of the fit is demonstrated in Fig. 5.23 by the recalculation of $R(\nu)$ which has to be compared with Fig. 5.21.

We have mentioned before the unusually strong phonon at 152 cm$^{-1}$. It is about four times larger in its oscillator strength as expected from lattice dynamical calculations. There is no reasonable way to explain this by a change of local charges on the ions. We think that the large strength is caused by a charge transfer during the vibration from the apical oxygen to the $CuO_2$ planes [5.51].

### 5.3.7 Gap Determination on $YBa_2Cu_3O_7$-Oriented Films

Here we describe a possible multi-reflection method for getting information about the very low absorption due to the strong reflection on the high-$T_c$ cuprates at low $T$, see also Appendix 5.AII, and $\nu$ [5.52]. The system is

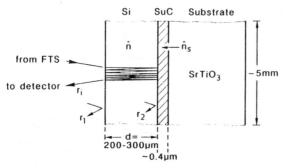

**Fig. 5.24.** Scheme of the Si reflection Fabry-Pérot (Si-RFP) for the multi-reflection analysis of high-$T_c$ superconductors

schematically presented in Fig. 5.24. There one has an oriented film, here of $YBa_2Cu_3O_7$, with the $c$-axis perpendicular to the plane on a substrate, such as $SrTiO_3$. The film should be opaque (thickness $\approx 0.4\,\mu m$) for reasons explained in Sect. 5.3.4. This sandwich is then pressed onto a Si wafer of a thickness of about 200–300 $\mu m$. It should be very pure and highly plane parallel (better than a few arc sec) and the superconducting film should be correspondingly plane (to a small fraction of a $\mu m$). From the Si side we can consider the system as a moderately efficient reflection Fabry-Pérot (RFP). One reflector is the superconductor, the other is the Si surface ($R \simeq 30\%$). The Si-RFP is then placed as a reflection sample into a FT spectrometer of resolution of at least $0.1\,cm^{-1}$ in order to resolve the RFP resonances completely. The FTS avoids the overlapping order problem which normally occurs with the use of Fabry-Pérot interferometers. The whole system also maintains the throughput advantage and partly the multiplex advantage of FTS. An Au mirror replaces the Si-RFP as the reference mirror.

A measurement with a rather good $YBa_2Cu_3O_7$-film (plasma frequency $\approx 12\,000\,cm^{-1}$, scattering rate $\approx 27\,cm^{-1}$ for 10 K and 70 $cm^{-1}$ for 100 K) is shown in Fig. 5.25 for $T = 10$ K and 100 K in the spectral range from $50\,cm^{-1}$ to $650\,cm^{-1}$. The region above $550\,cm^{-1}$ is really of no interest since an (in itself weak) absorption due to multiphonon effects in Si sets in. It is shown in [5.52] that the effective number of reflections in the center of the resonances is between 10 and 15, which is also the factor of merit for absorption compared to a single reflection on the superconductor. In principle, one could increase this factor by a thicker Si wafer but then one gets difficulties with the resolution in the FTS. As also explained in detail in [5.52], one best considers the absorption $A_0(= 1 - R_0)$ in the center of each resonance and takes the ratio of $A_0$ (10 K) to $A_0$ (100 K) (Appendix 5.A2), which is shown in Fig. 5.26. Below $50\,cm^{-1}$ one finds $A_0$ (10 K) so small and the signal too noisy for proceeding further to lower $\nu$. One calculates for the single reflection on the superconductor at $50\,cm^{-1}$ and 10 K a value of 99.97%! The further analysis assumes that the Drude absorption can be used above $T_c$ and that the electrodynamic

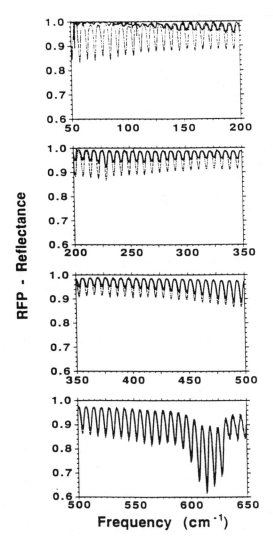

**Fig. 5.25.** The Si-RFP resonance spectra of an oriented film (c-axis perpendicular to surface) of $YBa_2Cu_3O_7$ on $SrTiO_3$ at 10 K (*full line*) and 100 K (*dotted line*). Resolution of FTS was $0.1\,cm^{-1}$. Thickness: Si wafer 217 μm, superconductor $\approx 0.4\,\mu m$

response theory given in [5.53, 54], see also Appendix 5.AIII, is appropriate for the superconducting state below $T_c$. In the latter, enter possible gap frequencies $\nu_{gj}$ and their weights $f_{gj}(\sum_j f_{gj} = 1)$ which are some measure of the relative area which the gaps take on the Fermi surface. There also enters into the response theory the ratio of the scattering rate $\nu_\tau$ and the gap frequency $\nu_g$ as parameters allowing one to cover the region between the clean and the dirty limit.

In Fig. 5.26 it is demonstrated that one or two gaps are unable to explain the result. It is furthermore shown that above $400\,cm^{-1}$, gaps can no longer

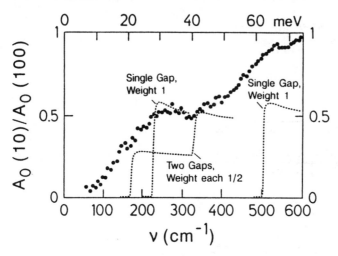

**Fig. 5.26.** The absorption ratio $A_0$ (10 K)$/A_0$ (100 K) vs frequency of the RFP resonances of Fig. 5.25 in their respective maximum position. Shown also are three calculated curves for a single gap or two gaps, demonstrating the impossibility of explaining the $A_0$-ratio in such a simple manner

be solely responsible for the $A_0(10)/A_0(100)$ ratio found. We analyzed the curve, therefore, by starting from low frequencies with a series of close-lying discrete gaps and appropriate weights, see Appendices 5.AII, AIII). Also some additional information had to be taken into account. Already earlier measurements on ceramics indicated the existence of a gap around 320–330 cm$^{-1}$. Raman measurements [5.55] especially confirmed this strongly by using a theoretical prediction [5.56] concerning the position and width of phonons below or above such a gap. We therefore had to place a gap bunch there. Furthermore, band structure calculations [5.57] predicted various interband transitions in the 200–600 cm$^{-1}$ region and we found such a broad band between 400 and 600 cm$^{-1}$ on oxygen reduced ceramic samples [5.58] (Fig. 5.27), showing an exotic temperature dependence of its integrated strength. It went down rapidly with increasing temperature by going to $T_c$ in such a way that the superconducting order parameter is reduced. This might indicate that the Cooper pairs at low $T$ are not very effective in screening the dipolar fields of the electronic band while the quasiparticles created at higher $T$ due to pair-breaking can do it. A similar band, as mentioned above, has been found in $Tl_2Ba_2Cu_3O_{10}$ [5.59]. Finally, we had to take into account the already mentioned condition $\sum_j f_{gj} = 1$ which limited the possibility of adding too many gaps.

In this way we got the fit for Fig. 5.26, as shown in Fig. 5.28. This fit is supported by further measurements of $A_0(T)/A_0(100)$ for $T = 70$ K and 85 K in addition to the $A_0(10)/A_0(100)$ result (Fig. 5.29). There are uncertainties for the fit of these latter data since one has to make assumptions about the

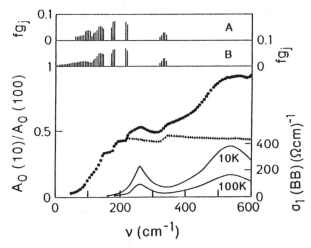

**Fig. 5.27.** Fit of the $A_0(10)/A_0(100)$ curves of Fig. 5.26 with two options $A$ and $B$ for a discrete multigap distribution with position $\nu_{gj}$ and weight $f_{gj}(\sum_j f_{gj} = 1)$. The fit also take into account a fit of the single resonances for 10 K and 100 K of Fig. 5.25, and also two interband transitions shown in the figure. The points marked with open circles represent the fit from the gaps alone. Explanations in the text

temperature dependencies of the gap positions, the scattering rate, and the assumed approximation of a two-fluid model. We have used with $t = T/T_c$

$$\nu_{gj}(t) = \nu_{gj}(0)\sqrt{1 - t^4}, \tag{5.37}$$

$$\nu_r(t) = 27 + 37.5t^4 \quad \text{for} \quad t < 1, \qquad \nu_r(t) = 64.5t \quad \text{for} \quad t \geq 1, \tag{5.38}$$

$$\hat{\sigma}(\nu, t) = (f_n + (1 - f_n)t^4)\hat{\sigma}_n(\nu, t) + (1 - f_n)(1 - t^4)\hat{\sigma}_s(\nu, t) + \sigma_{BB}(\nu, t). \tag{5.39}$$

$\hat{\sigma}_n$ is the Drude conductivity of the quasi-particles at $t$ and $\hat{\sigma}_s$ the superconducting dynamical conductivity, as mentioned before. $f_n$ denotes a small fraction of unpaired carriers which might exist even at $t = 0$. An analysis of all data of the microwave region, so far puplished, indicate that even the best samples have apparently $f_n$-values of a few percent. The origin is not clear.

The fit of Fig. 5.28 is of course not unique. This is especially true since we have experimental information only down to 50 cm$^{-1}$. We, therefore, consider two options for the gap distribution on the energy scale. The option $A$ of Fig. 5.28 ranges the gaps with their lower frequency part from 50 cm$^{-1}$ to 220 cm$^{-1}$, to option B, instead, from 5 cm$^{-1}$ to 220 cm$^{-1}$. In both cases the discrete gap distribution, only applied for an easier computation, is really thought to be a continuous distribution. In order to fit the $A_0(10)/A_0(100)$ ratio near 50 cm$^{-1}$, we roughly use for the option $A$ an $f_n$ value of about 4–7%, while the option $B$ needs only an $f_n$ between 0% and 2%. A decision

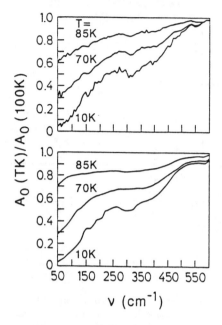

**Fig. 5.28.** Measurement (*upper part*) and fit (*lower part*) of the absorption ratio $A_0(T\,K)/A_0(100\,K)$ for $T = 10\,K$, $70\,K$ and $85\,K$, respectively, for the same film used for Figs. 5.25, 26

between the two distributions $A$ and $B$ cannot be made from our measurements. The same is true concerning the question about the pairing symmetry, i.e., about $s$ or $d$-pairing. The distribution $B$ might look as appropriate for $d$-wave pairing. However, $s$-wave pairing can also result in such an anisotropy of the gap. The FIR measurements in general cannot directly map the gaps on the Fermi surface. One could, perhaps, speculate whether our highest gaps around $330\,cm^{-1}$ are located near the saddle points [5.50] on the Fermi surface since there are places of high density of states. Another possible way of obtaining some information about this question is doping, for instance Fe-doping. It is known that Fe predominantly replaces Cu in the chains of

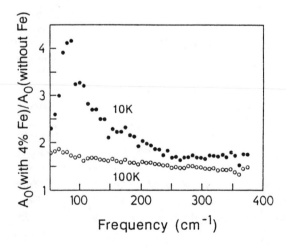

**Fig. 5.29.** Measurement of an iron-doped film of $YBa_2(Cu_{0.96}F_{0.04})_3O_7$ with the Si-RFP. Plotted is the absorption ratio $A_0$ (with Fe)/$A_0$ (without Fe) for $10\,K$ and $100\,K$, respectively. Explanation in text

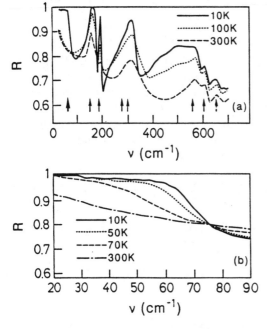

**Fig. 5.30 a,b.** Reflectance measurements on a 4% iron-doped ceramic showing in (**a**) phonons (*weak arrows*), a gap at 10 K (*strong arrow*) and a broad band between 400 and 600 cm$^{-1}$ with an exotic temperature dependence of its strength. In (**b**) the gap at about 60 cm$^{-1}$ for 10 K is shown and its change with temperature

YBa$_2$Cu$_3$O$_7$. An oriented film of YBa$_2$Cu$_{0.96}$Fe$_{0.04}$)$_3$O$_7$ gave the result shown in Fig. 5.30, which indicates rather clearly a strong change of the gap distribution just above 60 cm$^{-1}$ at 10 K. This is completely in accordance with an earler direct reflection measurement on a ceramic sample with the same iron concentration [5.58] (Fig. 5.27). The conclusion is, therefore, that the lower

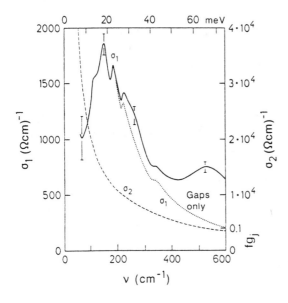

**Fig. 5.31.** The dynamic conductivities $\sigma_1(\nu)$ and $\sigma_2(\nu)$ for 10 K and between 50 and 600 cm$^{-1}$ of the YBa$_2$Cu$_3$O$_7$ film from Figs. 5.25–27

part of our gap distribution is connected with the chain band on the Fermi surface.

Finally, we would like to show the result of our analysis on the undoped material for the dynamical conductivity $\sigma_1$ and $\sigma_2$ in the FIR at 10 K based on the Figs. 5.25, 28, see also Fig. 5.31. The distributions $A$ and $B$ give here the same spectra within the experimental accuracy. Also shown is $\sigma_1(\nu)$ resulting from the gaps alone. $\sigma_2(\nu)$ is proportional to $\nu^{-1}$ in the whole range up to 400 cm$^{-1}$, since it represents the Kramers-Kronig transform of the supercurrent $\delta$-function at frequency zero. $\sigma_2$ is somewhat dependent on the gap distribution and on the scattering rate $\nu_r$. For $t = T/T_c < 1$ and $\omega \ll \omega_{gj}(t)$ it can be expressed quite accurately (in cgs units) by

$$\sigma_2 = a_\tau \frac{1}{4\pi} \frac{\omega_p^2}{\omega}; \quad a_\tau = (1 - f_n)(1 - t^4) \sum_j f_{gj} \left(1 + \frac{2}{\pi} \frac{\omega_r(t)}{\omega_{gj}(t)}\right). \quad (5.40)$$

The $a_r$-formula follows from Appendix 5.AIII (5.A27) and from (5.39). It holds for the whole region from the clean to the dirty limit.

### 5.3.8 Extrapolation to Microwaves with Results from Sect. 5.3.7

In the microwave spectral region two quantities are of prime importance, namely the field penetration depth $\lambda$, and the complex surface impedance $\hat{Z}_s = R_s + iX_s$ with $R_s$, being the surface resistance and $X_s$ the surface reactance [5.60]. $\lambda$ is generally given by the reciprocal of the imaginary part of the light wavevector $\hat{k}_L$ in the material

$$\lambda = (\text{Im } \hat{k}_L)^{-1} = \frac{c}{\omega\kappa} = \frac{1}{2\pi\nu\kappa}, \quad (5.41)$$

where $\kappa$ is the imaginary part of the complex refractive index $\hat{n} = n + i\kappa$; $c$ is the vacuum light velocity, and $\omega$ the angular frequency. For $t = T/T_c < 1$ we can express $\kappa$ with the help of the $a_\tau$ function of (5.40) by

$$\kappa = \sqrt{a_\tau} \frac{\nu_p}{\nu}. \quad (5.42)$$

In the extreme clean limit of a superconductor, the London limit, it holds $\nu_\tau \ll \nu_g$, or the coherence length $\xi_0$ is small compared to the mean free path $\ell$ of the carriers. In this limit one has $a_\tau = 1$ and $\nu\kappa = \nu_p$ the unscreened plasma frequency, and, therefore,

$$\lambda_L = \frac{c}{\omega_p} = \frac{1}{2\pi\nu_p}, \quad (5.43)$$

with

$$\omega_p = \frac{4\pi N e^2}{m^*} \quad (5.44)$$

in obvious notation. Combining (5.41) with (5.42, 43) we get the simple and useful relations

$$\lambda = \frac{\lambda_{\mathrm{L}}}{\sqrt{a_\tau}}, \quad \lambda^2 = \frac{c^2}{4\pi\omega}\frac{1}{\sigma_2}. \tag{5.45}$$

The surface impedance can be expressed by

$$\hat{Z}_{\mathrm{s}} = \frac{R_{\mathrm{v}}}{n - \mathrm{i}\kappa}, \quad R_{\mathrm{v}} = 376.7\,\Omega. \tag{5.46}$$

We have then in the microwave region for $T < T_{\mathrm{c}}$ $(\kappa^2 \gg n^2)$:

$$R_{\mathrm{s}} = R_{\mathrm{v}}\frac{n}{n^2 + \kappa^2} \simeq R_{\mathrm{v}}\frac{n}{\kappa^2}, \tag{5.47}$$

$$X_{\mathrm{s}} = R_{\mathrm{v}}\frac{\kappa}{n^2 + \kappa^2} \simeq R_{\mathrm{v}}\frac{1}{\kappa} = R_{\mathrm{v}}2\pi\nu\lambda. \tag{5.48}$$

In the London limit follows finally $X_{\mathrm{s}} = R_{\mathrm{v}}\nu/\nu_{\mathrm{p}}$.

It is, of course, dangerous to make extrapolations from the FIR to the microwave region. We will nevertheless do it in order to challenge for tests. We first look at the $t$ dependence of the penetration depth $\lambda$ in the superconducting state of the YBa$_2$Cu$_3$O$_7$ film $(E \perp c)$ with $\nu_{\mathrm{p}} = 12\,000\,\mathrm{cm}^{-1}$ and for $\nu < 90\,\mathrm{GHz}$ $(3\,\mathrm{cm}^{-1})$. Most critical is the temperature dependence of the scattering rate $\nu_\tau$ (5.38). We choose three cases:

a)  $\nu_\tau = 27 + 110\,t$      $[\mathrm{cm}^{-1}]$;

b)  $\nu_\tau = 27 + 210\,t^2$      $[\mathrm{cm}^{-1}]$;

c)  $\nu_\tau = 27 + 37.5\,t^4$      $[\mathrm{cm}^{-1}]$,

where the case c) is closest to our result of Fig. 5.29. Figure 5.32 gives the result for $\lambda$. It shows that $\lambda(t)$ for $t < 0.2$ follows directly the $t$ dependence of $\nu_\tau$. The same is true for the quantity $1 - \lambda^2(0)/\lambda^2(t)$ as shown in Fig. 5.33. This quantity and $\lambda(t) - \lambda(0)$ have been considered in their behavior at lower $t$ as a test for $s$- or $d$-wave pairing [5.61, 62]. How clear this test is in view of our results might be a question of debate. It should be noted in this context that the $\nu_\tau(t)$ of cases a and b has been chosen so that the quantities $\lambda^2(0)/\lambda^2(t)$ and $\lambda(t) - \lambda(0)$ found in [5.62] and [5.61], respectively, are described quite well.

Next, we look at the $t$ dependence of the surface resistance $R_{\mathrm{s}}$ and the surface reactance $X_{\mathrm{s}}$, Fig. 5.34. While $X_{\mathrm{s}}$ is almost independent of $t$ up to about $t = 0.5$, one observes, as is well-known, a strong $t$ and $\nu$-dependence of $R_{\mathrm{s}}$, and also quite a large influence on $f_n$ for the lowest temperatures. Finally, the frequency dependence of $R_{\mathrm{s}}$ and $X_{\mathrm{s}}$ is considered in Fig. 5.35 for $T = 10\,\mathrm{K}$ and the gap distribution $A$ and $B$ with some $f_n$-values.

As expected from (5.45, 48), one finds $X_{\mathrm{s}} \propto \nu$ and independent of $f_n$. $R_{\mathrm{s}}$ is proportional to $\nu^2$ below $10\,\mathrm{cm}^{-1}$, as has also been found experimentally [5.63]. Furthermore, the plot shows, as mentioned above, that we cannot decide from the FIR measurements, down to only $50\,\mathrm{cm}^{-1}$, whether the gap

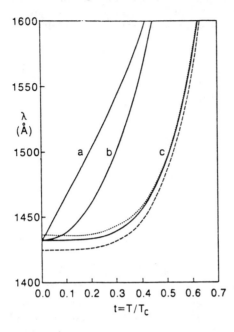

**Fig. 5.32.** Extrapolation to microwaves: Field penetration depth $\lambda$ vs $t = T/T_c$ for frequencies below 90 GHz of an YBa$_2$Cu$_3$O$_7$ film with $E \perp c$ and $\nu_p = 12\,000\,\text{cm}^{-1}$. The curves $a$, $b$, and $c$ have different scattering rates $\nu_r(t)$ given in the text. *Full curves*: The $B$ distribution of Fig. 5.28 with $f_n = 0.01$. *Dashed curve*: The $B$ distribution with $f_n = 0$. *Dotted curve*: The $A$ distribution with $f_n = 0.04$. For low $t$ the $\lambda(t)$ curves follow the $\nu_r(t)$ dependence

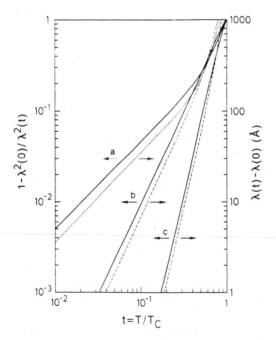

**Fig. 5.33.** Extrapolation to microwaves: The quantities $\lambda(t) - \lambda(0)$ and $1 - \lambda^2(0)/\lambda^2(t)$ vs $t$ for $\nu = 3\,\text{GHz}$, and for the three cases of $\nu_r(t)$ as in Fig. 5.32 with $f_n = 0$. The $\lambda(0)$ is 1424 Å

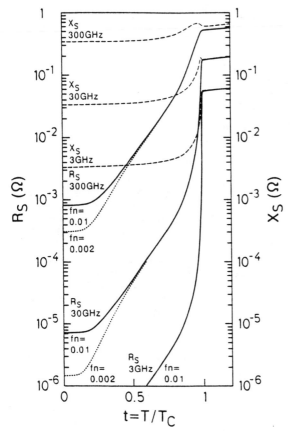

**Fig. 5.34.** Extrapolation to microwaves: Surface resistance $R_s$ and reactance $X_s$ vs $t$ for 3 GHz, 30 GHz and 300 GHz, respectively. The data are for a $YBa_2Cu_3O_7$ film, $E \perp c$, with $\nu_p = 12\,000\,cm^{-1}$, $T_c = 92\,K$ and for $\nu_r = 27 + 37.5t^4$ $(t < 1)$ and $\nu_r = 64.5t$ $(t > 1)$. The strong influence of $f_n$ at low $t$ for $R_s$ is shown

anisotropy $A$ with $f_n \sim 0.07$ or the anisotropy $B$ with $f_n \sim 0.01$ should be taken. A few lower experimental points in Fig. 5.35 between 55 cm$^{-1}$ and 85 cm$^{-1}$ demonstrate this.

Finally, we make an extrapolation of the conductivity $\sigma_1(t, f_n)$ to microwave frequencies 100 GHz. This is shown in Fig. 5.36 which holds for both anisotropies given in Fig. 5.28. The curves $\sigma_1(t, f_n)$ are based on the (5.37–5.39). The dissipation is caused here for $f_n = 0$, solely by quasiparticles from thermally broken Cooper pairs. For low $t$ one finds a slope of $\sigma_1$ proportional to $t^4$, according to the use of the two-fluid model, (5.39). This is, of course, a question of an experimental test. The same holds also for the large influence of the $f_n$ value below $t = 0.1$, since it is assumed that the non-paired quasi-particles, described by $f_n$, behave in a Drude-like manner.

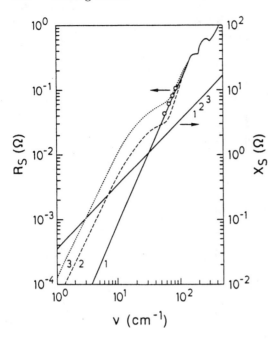

**Fig. 5.35.** Extrapolation to microwaves: Surface resistance $R_s$ and reactance $X_s$ vs $\nu$ at $10\,\mathrm{K}$ for a $YBa_2Cu_3O_7$ film, $E \perp c$, with $\nu_p = 12\,000\,\mathrm{cm}^{-1}$ and $\nu_r = 27\,\mathrm{cm}^{-1}$. *Curve 1:* $B$ distribution with $f_n = 0.01$. *Curve 2:* $A$ distribution with $f_n = 0.07$. *Curve 3:* $A$ distribution with $f_n = 0.14$

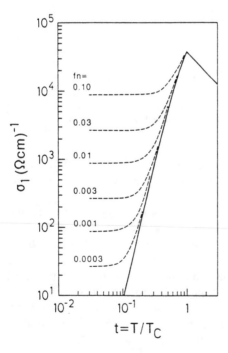

**Fig. 5.36.** Extrapolation to microwaves: $\sigma_1(\nu)$ of a $YBa_2Cu_3O_7$ film, $E \perp c$, for $\nu < 120\,\mathrm{GHz}$ $[4\,\mathrm{cm}^{-1}]$ and for $t = T/T_c$ from 0.03 to 3. Calculation is based on (5.39). The parameter is $f_n$. No coherence effect near $t = 1$ in the calculation

### 5.3.9 Far Infrared Region – Fourier Transform Spectroscopy with Ellipsometry on $La_{1.87}Sr_{0.13}CuO_4$

Only recently the ellipsometry was extended to the IR [5.64–66] and to the FIR [5.30, 31], and combined with FTS. It is shown in this section that such a rotating analyzer ellipsometry attached to an FT spectrometer (Bruker IFS-113v) can analyze the dielectric tensor of the highly anisotropic crystalline material of the superconductor $La_{1.87}Sr_{0.13}CuO_4$ in the spectral range from $30\,cm^{-1}$ to $700\,cm^{-1}$ and down to 5 K [5.67]. Since ellipsometry directly yields the amplitude and phase information one overcomes the problems introduced by a Kramers-Kronig analysis at low frequencies and by reference mirrors. On the other hand, since one has to use a strongly oblique incidence (here 80°) for such a superconductor, the sample has to be rather large. The experiment was done on a mosaic of several single crystals which were cut from a larger crystal giving a useful sample area of $8 \times 8\,mm^2$. The uniaxial sample was oriented with the surface being (100). The measurement yields the complex reflection ratio $\hat{\varsigma} = r_p/r_s$ ($c$-axis parallel and perpendicular to the plane of incidence), and the analysis is done by fitting with the Fresnel equations. Of interest was the dielectric function $\varepsilon_c = \varepsilon_{1c} + i\varepsilon_{2c}$ of the $c$ direction concerning the electronic and phononic contributions. The results are shown in Fig. 5.37a,b for 5 K and 300 K (RT). There are five phonons with frequencies (for 5 K) at $234.9\,cm^{-1}$, $250.5\,cm^{-1}$, $312\,cm^{-1}$, $353\,cm^{-1}$, and $491.5\,cm^{-1}$, respectively. The lowest phonon has an anomalously high oscillator strength (5.36), about 7 times larger than standard shell model calculations would give, assuming local ionic charges. This again indicates a charge-transfer mechanism as it has been found in $YBa_2Cu_3O_7$ for $E\|c$ [5.49, 51]. Below $100\,cm^{-1}$ and at 5 K, one finds characteristic effects for superconductivity ($T_c$ was 31 K) and a fit can be done with a single gap at about $55\,cm^{-1}$ and an unscreened plasma frequency of only $400\,cm^{-1}$. See Sect. 5.3.6 for the corresponding case of $YBa_2Cu_3O_7$. The experiment failed, however, to observe the expected upturn below about $60\,cm^{-1}$ at 300 K due to free carrier absorption.

Recently, it has been shown that one also can get the $c$-axis response of a high-$T_c$ thin film or crystal ($Tl_2Ba_2Ca_2Cu_3O_{10}$), [5.68]) which is oriented with the $c$-axis perpendicular to the surface by performing polarized reflectivity measurements under oblique incidence. Ellipsometry was not used, instead the reflectance $R_s(\omega)$ and $R_p, (\omega)$ for $s$- and $p$-polarized light, respectively, has been measured and analyzed by KK-transformations.

# Appendix

### 5.AI: Kramers–Kronig Analysis

The Kramers-Kronig (KK) integral transformation generally connects the real and imaginary parts of a response quantity such as $\varepsilon_1(\omega)$ and $\varepsilon_2(\omega)$ of

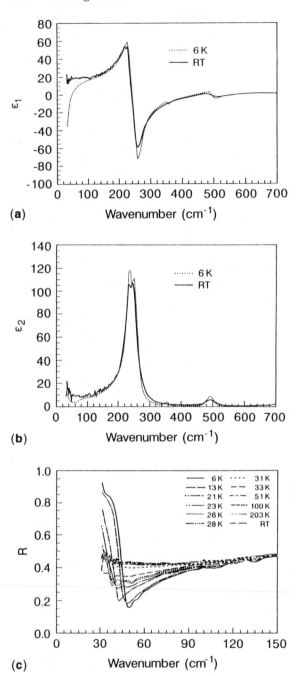

**Fig. 5.37 a–c.** $\varepsilon_1(\nu)$ and $\varepsilon_2(\nu)$ of a $La_{1.87}Sr_{0.13}CuO_4$ single crystal, $c$ direction at 5 K and 300 K (RT). The measurements were done with FIR-FTS combined with ellipsometry. The feature occuring at $\approx 185\,cm^{-1}$ is an experimental artifact

the dielectric function of a solid. It can only be applied for a linear response. It is

$$\varepsilon_1(\omega) - \varepsilon_\infty = \frac{2}{\pi} P \int_0^\infty \frac{\omega' \varepsilon_2(\omega')}{\omega'^2 - \omega^2} d\omega', \tag{5.A1}$$

$$\varepsilon_2(\omega)_\infty = -\frac{2\omega}{\pi} P \int_0^\infty \frac{\varepsilon_1(\omega')}{\omega'^2 - \omega^2} d\omega'. \tag{5.A2}$$

The symbol $P$ denotes Cauchy's principle part of the integral. $\varepsilon_\infty$ plays no role under the integral (5.A2), since generally for $\omega \neq 0$ one has

$$P \int_0^\infty \frac{1}{\omega'^2 - \omega^2} d\omega' = 0. \tag{5.A3}$$

In case of a finite conductivity of free carriers, one has a divergence of $\varepsilon_2$ at $\omega = 0$. (5.A2) here has to be changed to

$$\varepsilon_2(\omega) - \frac{4\pi\sigma_0}{\omega} = -\frac{2\omega}{\pi} P \int_0^\infty \frac{\varepsilon_1(\omega')}{\omega'^2 - \omega^2} d\omega', \tag{5.A4}$$

with $\sigma_0$ being the d.c. conductivity (cgs units). This contribution has no influence on $\varepsilon_1$ in (5.A1).

For superconductors at $T < T_c$ one finds that $\varepsilon_1$ diverges below gaps as $\omega^{-2}$ to $-\infty$. Then (5.A1) has to be changed to

$$\varepsilon_1(\omega) - \frac{8A}{\omega^2} - \varepsilon_\infty = \frac{2}{\pi} P \int_0^\infty \frac{\omega' \varepsilon_2(\omega')}{\omega'^2 - \omega^2} d\omega', \tag{5.A5}$$

where $A$ is the strength of the $\delta$-function of $\sigma_1$ at frequency zero:

$$A = a_\tau \frac{\omega_p^2}{8}, \tag{5.A6}$$

with $a_\tau$ being the function given in (5.40).

In most cases the KK transformation is applied if the reflectance $R(\omega)$ of a sample has been measured over a large spectral range. In this case the analysis determines the reflection amplitude $r(\omega)$ and the reflection phase $\phi_r(\omega)$ defined by or by

$$r(\omega) = \sqrt{R(\omega)} e^{i\phi_r(\omega)}, \tag{5.A7}$$

or by

$$\ln r(\omega) = \ln \sqrt{R(\omega)} + i\phi_r(\omega). \tag{5.A8}$$

One has, in this case, an analogy to (5.A2):

$$\phi_r(\omega) = -\frac{2\omega}{\pi}P\int_0^\infty \frac{\ln\sqrt{R(\omega')}}{\omega'^2 - \omega^2}d\omega'. \tag{5.A9}$$

In contrast to $\varepsilon_2$, however, $\phi_r$ is a gauge-dependent quantity. For instance, $r(\hat{\omega})$ can be multiplied with any function of $\hat{\omega}$, which has the absolute value of 1 on the real axis without changing the reflectance. In (5.A9), $\phi_r(\omega)$ can only have values between zero and $\pi$ without getting negative values of $\varepsilon_2$. The numerical calculation of (5.A9) can be simplified by subtracting a term from the integral of (5.A9) which gives no contribution to the integral. If (5.A9) is changed to

$$\phi_r(\omega) = -\frac{\omega}{\pi}P\int_0^\infty \frac{\ln R(\omega') - \ln R(\omega)}{\omega'^2 - \omega^2}d\omega', \tag{5.A10}$$

then the singularity at $\omega = \omega'$ of the Cauchy principle value is omitted. The complex refractive index is then calculated via (5.A7) and $r(\omega) = (n + i\kappa - 1)/(n + i\kappa + 1)$ by

$$n(\omega) = \frac{1}{N_e}[1 - R(\omega)]; \quad \kappa = \frac{1}{N_e}2\sqrt{R(\omega)}\sin\phi_r;$$

$$N_e = 1 + R(\omega) - 2\sqrt{R(\omega)}\cos\phi_r. \tag{5.A11}$$

(A detailed discussion of errors which occur with the KK analysis is found in [5.49]).

## 5.AII: Formulas for Si-Reflection Fabry–Pérot

We assume that the superconducting film (Fig. 5.24) is opaque and that one has a nearly vertical incidence of a collimated beam on it. By summing up the amplitudes inside and outside of the RFP (partial wave analysis) one finds:

$$r_i = \sqrt{R_i}e^{i\phi_i} = \frac{r_1 + ar_2}{1 + ar_1r_2}, \tag{5.A12}$$

where

$$a = |a|e^{i\phi_a} = e^{i4\pi\nu\hat{n}d},$$

with $\hat{n} = n + i\kappa$ being the complex refractive index of Si and $d$ its thickness. Furthermore, we define

$$r_1 = \sqrt{R_1}e^{i\phi_1}, \quad \phi_1 \simeq -\pi, \quad r_2 = \sqrt{R_2}e^{i\phi_2}. \tag{5.A13}$$

One then takes $R_i = |r_i|^2$, and after some algebra, $A_i = 1 - R_i$;

$$A_i(\nu) = \frac{A_0}{1 + \frac{4}{\pi^2}F^2\sin^2\left[\frac{1}{2}(\phi_a + \phi_2)\right]}, \tag{5.A14}$$

with

$$A_0 = \frac{(1 - R_1)(1 - |a|^2 R_2)}{(1 - |a|\sqrt{R_1 R_2})^2} \tag{5.A15}$$

being the maximal absorption in the center of the RFP resonance. $F$ denotes the finesse

$$F = \frac{\pi \sqrt[4]{|a|^2 R_1 R_2}}{1 - \sqrt{|a|^2 R_1 R_2}}, \tag{5.A16}$$

which is a measure for the number of effective reflections averaged over one resonance. $F$ is related to the resonance full half width by $\Delta\nu = \nu_s/(sF)$ where $s = 0, 1, 2, 3 \ldots$ denotes the number of the resonance of frequency $\nu_s$, which is determined by

$$\phi_a + \phi_2 = s \cdot 2\pi, \quad \phi_2(\nu_s) = -4\pi\nu_s nd + s \cdot 2\pi, \tag{5.A17}$$

thus

$$\nu_s = \frac{1}{2nd}\left(s - \frac{1}{2\pi}\phi_2\right) \simeq \frac{1}{2nd}\left(s + \frac{1}{2}\right). \tag{5.A18}$$

With $R_0 = 1 - A_0$ one finds from (5.A15) for $R_2(\nu_s)$

$$|a|\sqrt{R_2} = \frac{\sqrt{R_0} + \sqrt{R_1}}{1 + \sqrt{R_0 R_1}}, \tag{5.A19}$$

showing that $R_2(\nu_s)$ is determined if $|a|$ and $R_1$ of the Si wafer are known, and $R_0$ is measured. Furthermore, if $\nu_s$ is measured one knows $\phi_2(\nu_s)$ from (5.A18), thus giving $r_2(\nu_s)$ of (5.A13), and with this the complex refractive index of the superconductor:

$$\hat{n}_s(\nu_s) = \hat{n}\frac{1 - r_2(\nu_s)}{1 + r_2(\nu_s)}, \tag{5.A20}$$

which also yields the complex dielectric function $\hat{\varepsilon}_s = \hat{n}_s^2$. The real experimental problem is, however, to determine $\nu_s$ with high enough accuracy. But here it helps that the RFP phase $\phi_i$ goes through zero at the $\nu_s$ resonance frequencies. If the FTS has first roughly determined $\nu_s$, then one can correct $\nu_s$ with (5.A18) on the computer until $\phi_i = 0$ is reached. But the procedure is still somewhat problematic since the value of $R_0$, occurring in (5.A19), is also not well enough known. The reason for this is the incomplete collimation of the beam and the nonvertical incidence of the beam on the sample (Fig. 5.24). These last effects drop out, however, in the absorption ratio $A_0(10\,\mathrm{K})/A_0(100\,\mathrm{K})$, shown in Fig. 5.26, because they are merely of geometrical nature. The most straightforward way for getting the conductivity $\sigma_1$ at low $T$, (for instance, $10\,\mathrm{K}$), is the use of the empirical relation

$$\sigma_1(10\,\mathrm{K}) \simeq \beta\,\sigma_1(100\,\mathrm{K}) \cdot [A_0(10\,\mathrm{K})/_0(100\,\mathrm{K})]_{\mathrm{exp}}, \tag{5.A21}$$

where the factor $\beta$ is about 0.92 to 0.94. $\sigma_1(100\,\mathrm{K})$ can be reasonably well calculated from the Drude dielectric function since one is above $T_c$. If the approach of (5.A21) is used, one does not need to primarily make a fit with a gap distribution as applied in Sect. 5.3.7. Here this would be a secondary step in order to better understand the $\sigma_1(\nu)$ function at low $T$.

## 5.AIII: Electrodynamic Response of Superconductors

The response theory applicable to the whole region from the clean to the dirty limit is given in [5.53, 54]. Here we present a good analytical approximation [5.69] which facilitates the discussion, especially in the lower frequency range at the microwaves. We formulate the response with the dynamical complex conductivity

$$\hat{\sigma}_s = \sigma_0(S_1 + iS_2), \qquad (5.A22)$$

where $\sigma_0$ is the d.c. normal conductivity ($\sigma_0 = 0.016678\, \nu_p^2/\nu_\tau\, \Omega^{-1}\,\mathrm{cm}^{-1}$) and $\nu_\tau = \nu_\tau(T)$ the scattering rate. We introduce two parameters

$$X = \frac{\nu}{\nu_g}, \quad Y = \frac{\nu_\tau}{\nu_g} \qquad (5.A23)$$

with $\nu_g$ being the gap frequency. The approximation is then written as:

$$\text{For } X \leq 1: \quad S_1 = 0, \quad S_2 = X^{-1}[a - (a - b)X^2],$$

$$\text{For } X > 1: \quad S_1 = \frac{\sigma_{1n}}{\sigma_0}\left[1 - \left(\frac{1-V}{X-V}\right)^{\pi/2}\right]\left(1 + \frac{U}{X}\right), \qquad (5.A24)$$

$$S_2 = \frac{\sigma_{2n}}{\sigma_0} + W\exp\left(-\frac{3(X-1)}{Z+0.25}\right)$$

$$- \left(W - b + \frac{Y}{1+Y^2}\right)\exp\left(-\frac{X-1}{Z+0.25}\right). \quad (5.A25)$$

The following abbreviations have been used:

$$\frac{\sigma_{1n}}{\sigma_0} = \frac{\nu_\tau^2}{\nu^2 + \nu_\tau^2}, \quad \frac{\sigma_{2n}}{\sigma_0} = \frac{\nu\nu_\tau}{\nu^2 + \nu_\tau^2};$$

$$a = \frac{Y}{1 + \frac{2}{\pi}Y}, \quad b = Z - Z^2 + Z^3, \quad Z = \frac{Y}{1+Y};$$

$$c = 1.652(1.12Z^4 - Z^3)\exp[-2(1-Z)^2];$$

$$U = \frac{1.25 + 2.5\,Y^3}{3.6 + Y^4}, \quad V = \frac{1 + UY}{1 + Y^2},$$

$$W = \left[c - \left(b - \frac{Y}{1+Y^2}\right)\exp\left(-\frac{1}{Z+0.25}\right)\right]$$

$$\times \left[\exp\left(-\frac{3}{Z+0.25}\right) - \exp\left(-\frac{1}{Z+0.25}\right)\right]^{-1}.$$

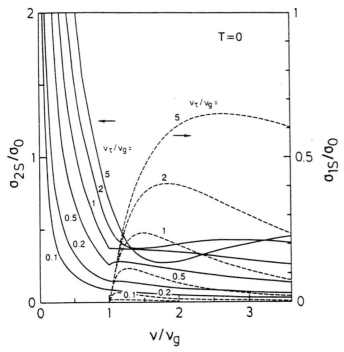

**Fig. 5.38.** $\sigma_{1s}$ and $\sigma_{2s}/\sigma_0$ as a function of the reduced frequency $\nu/\nu_g$ with $\nu_\tau/\nu_g$ as a parameter. The plot is according to (5.A24, A25) from the Appendix 5.AIII

Limiting cases for $T < T_c$:

$$X \gg 1 \quad : \quad S_1 = \frac{\sigma_{1n}}{\sigma_0}, \quad S_2 = \frac{\sigma_{2n}}{\sigma_0}; \tag{5.A26}$$

$$X \ll 1 \quad : \quad S_1 = 0, \quad S_2 = \frac{\nu_\tau}{\nu} \frac{1}{1 + \frac{2}{\pi}\frac{\nu_\tau}{\nu_g}}; \tag{5.A27}$$

clean limit:  $\nu_\tau \ll \nu_g$,  $S_2 = \frac{\nu_\tau}{\nu}$;
dirty limit:  $\nu_\tau \gg \nu_g$,  $S_2 = \frac{\pi}{2}\frac{\nu_g}{\nu}$.

If several gaps with frequencies $\nu_{gj}$ have to be taken into account, one then calculates with the gap weights $f_{gj}$

$$\hat{\sigma}_s = \sum_j f_{gj}\hat{\sigma}_{sj}, \tag{5.A28}$$

where

$$\sum_j f_{gj} = 1. \tag{5.A29}$$

It is worth noting that (5.A27) for $S_2$ is especially useful for the discussion of microwave data, see function $a_\tau$, (5.40) in Sects. 5.3.7, 8).

In Fig. 5.38 it is shown the function $S_1 = \sigma_{1s}/\sigma_0$ and $S_2 = \sigma_{2s}/\sigma_0$ as a function of $X = \nu/\nu_g$ according to (5.A24, A25). The parameter is $Y = \nu_\tau/\nu_g$. Note the dramatic reduction of $\sigma_{1s}/\sigma_0$ if $\nu_\tau/\nu_g$ becomes small, i.e., if one is going in the direction of the clean limit. This is one of the reasons for the difficulty in finding gaps as in the case of the high $T_c$ cuprates. It shows also that gaps at low frequencies yield a stronger onset of $\sigma_{1s}$ than those at higher frequencies, because the latter ones are closer to the clean limit. If $\nu_\tau$ would really go to zero then the FIR spectroscopy cannot detect gaps at all. The slope of all $\sigma_{1s}/\sigma_0$ curves at $\nu/\nu_g = 1$ is equal to $\pi/2$. For a general survey regarding the IR and FIR properties of the high-$T_c$ superconductors, the reader is referred to the literature [5.70, 71].

# References

5.1    J. Strong: *Concepts of Classical Optics* (Freeman, San Francisco 1958)

5.2    L. Mertz: *Transformation in Optics* (Wiley, New York 1965)

5.3    R. Bracewell: *The Fourier transform and its Applications* (McGraw-Hill, New York 1965)

5.4    W.H. Steel: *Interferometry* (Cambridge Univ. Press, New York 1967)
       G.A. Vanasse, A.T. Stair, D.J. Baker (eds.): Aspen Int'l Conf. on Fourier Spectroscopy (1970) p.43

5.5    D.H. Martin (ed.): *Spectroscopic Techniques for Far-Infrared, Submillimeter and Millimeter Waves* (North-Holland, Amsterdam 1967)

5.6    J.E. Stewart: *Infrared Spectroscopy* (Dekker, New York 1970)

5.7    G.W. Chantry: *Submillimetre Spectroscopy* (Academic, London 1971)

5.8    K.D. Möller, G. Rothschild: *Far-Infrared Spectroscopy* (Wiley, New York 1971)

5.9    R.J. Bell: *Introductory Fourier Transform Spectroscopy* (Academic, New York 1972)

5.10   D.C. Champeney: *Fourier Transforms and Their Physical Applications* (Academic, London 1973)

5.11   P.R. Griffiths: *Chemical Infrared Fourier Transform Spectroscopy* (Wiley, New York 1975)
       P.R. Griffiths, J.A. de Haseth: In *Fourier Transform Infrared Spectroscopy*, ed. by P.J. Elving, J.D. Winefordner, I.M. Kolthoff (Wiley, New York 1986)

5.12   K.D. Möller: *Optics* (University Sci. Books, Mill Valley, CA 1988)

5.13   P.B. Fellgett: J. Phys. Radium **19**, 197–237 (1958)

5.14   H. Rubens, R.W. Wood: Philos. Mag. **21**, 249 (1911)

5.15   A.A. Michelson: Philos. Mag. **31** (5), 256: 338 (1891)

5.16   J. Connes: J. Phys. Radium **19**, 197 (1958)

5.17   J. Strong, G.A. Vanasse: J. Opt. Soc. Am. **49**, 844 (1959)

5.18   P. Jacquinot: Rept. Progr. Phys. **13**, 267 (1960); Appl. Opt. **8**, 497 (1967)

5.19   J. Connes: Rev. Opt. **40**, 45, 116, 171, 231 (1961)

5.20   H.A. Gebbie: Adv. Quantum Electr. (Columbia Univ. Press, New York 1961) p.155; Appl. Opt. **8**, 501 (1967)

5.21   P.L. Richards: J. Opt. Soc. Am. **54**, 1474 (1964)

5.22   E.V. Loewenstein: Appl. Opt. **5**, 845 (1966); Aspen Int'l Conf. on Fourier Spectroscopy, ed. by G. Vanasse, A.T. Stair, D.J. Baker (1970) p.3

5.23   G.A. Vanasse, H. Sakai: *Progr. Optics* **6**, 261 (North-Holland, Amsterdam 1967)

5.24   L. Genzel: Fourierspektroskopie, in *Plenarvorträge der 33. Physikertagung* (1968) (Teubner, Stuttgart 1969) p.128; J. Mol. Spectrosc. **4**, 241 (1960); Fresenius Z. Anal. Chem. **273**, 391 (1975)

5.25   P. Grosse: Spektroskopie und Fouriertransformation, Beckman Rept. **1/70**, 3 (1970)

5.26   L. Mertz: Appl. Opt. **10**, 386 (1971)

5.27   R. Geick: *Topics in Current Chemistry* **58**, 75 ((Springer, Berlin, Heidelberg 1975); Fresenius Z. Anal. Chem. **288**, 1 (1977)

5.28   L. Genzel, K. Sakai: J. Opt. Soc Am. **67**, 871 (1977)

5.29   J.R. Birch, T.J. Parker: A Bibliography on Dispersive Fourier Transform Spectrometry. Infrare Phys. **19**, 201 (1979)

5.30   K.-L. Barth, D. Böhme, K. Kamaras, F. Keilmann, M. Cardona: Thin Solid Films **234**, 314 (1993)

5.31   K.-L. Barth, F. Keilmann: Rev. Sci. Instr. **64**, 870 (1993)

5.32   J.-W. Cooley, J.W. Tukey: Math. Comput. **19**, 297 (1965)

5.33   M.L. Forman: J. Opt. Soc. Am. **56**, 978 (1966)

5.34   L. Genzel, H.R. Chandrasekhar, J. Kuhl: Opt. Commun. **18**, 381 (1976)

5.35   L. Genzel, J. Kuhl: Infrared Phys. **18**, 113 (1978)

5.36   D.H. Martin, E. Puplett: Infrared Phys. **10**, 105 (1970)

5.37   D.H. Martin: In *Infrared and mm Waves*, ed. by K.J. Button (Academic, New York 1892) Chap.2

5.38   L. Genzel, A. Poglitsch, S. Haeseler: Int'l J. IR and MM Waves **6**, 741 (1985)

5.39   T.M. Lifshifts, F. Ya Nad': Sov. Phys. - Dok. **10**, 532 (1965)

5.40   Sh. M. Kogan, T.M. Lifshits: Phys. Stat. Solidi (a) **39**, 11 (1977)

5.41   M.S. Skolnick, L. Eaves, R.A. Stradling, J.C. Portal, S. Askenazy: Solid State Commun. **15**, 1403 (1974)

5.42   E.E. Haller, B. Joos, L. Falicov: Phys. Rev. B **21**, 4729 (1980)

5.43   J.A. Griffin: Photothermal ionisation spectroscopy studies on shallow impurities in silicon, Disseration, University of Stuttgart (1987)

5.44   S.C. Shen, C.J. Fang, M. Cardona, L. Genzel: Phys. Rev. B **22**, 2913 (1980)

5.45   J. Gast, L. Genzel: Opt. Commun. **8**, 26 (1973)

5.46    H. Bilz, D. Strauch, R.K. Wehner: *Encyclopedia of Physics* **XXV/2d**, ed. by S. Flügge, L. Genzel (Springer, Berlin, Heidelberg 1984)

5.47   C. Thomsen, M. Cardona, W. Kress, R. Liu, L. Genzel, M. Bauer, E. Schönherr, U. Schröder: Solid State Commun. **65**, 1139 (1988)

5.48   M. Bauer, I.B. Ferreira, L. Genzel, M. Cardona, P. Murugaraj, J. Maier: Solid State Commun. **72**, 551 (1989)

5.49   M. Bauer: Untersuchungen an Hochtemperatursupraleitern von Typ $YBa_2Cu_3O_{7-\delta}$ mit Hilfe der Infrarot-Spektroskopie. Dissertation, University of Tübingen (1990)
       K.F. Renk: In *Studies of High-Temperature Superconductors*, ed. by A.V. Narlikar (Nova Sci., New York 1992)

5.50   O.K. Andersen, A.I. Liechtenstein, C.O. Rodriguez, I.I. Mazin, O. Jepsen, V.P. Antropov, O. Gunnarson, S. Gopolan: Physica C **185-189**, 147 (1991)

5.51  L. Genzel, A. Wittlin, M. Bauer, M. Cardona, E. Schönherr, A. Simon: Phys. Rev. B **40**, 2170 (1989)

5.52  L. Genzel, M. Bauer, H.-U. Habermeier, E.H. Brandt: Z. Physik B **90**, 3 (1993); Solid State Commun. **81**, 589 (1992)

5.53  W. Lee, D. Rainer, W. Zimmermann: Physica C **159**, 535 (1989)

5.54  W. Zimmermann, E.H. Brandt, M. Bauer, E. Seider, L. Genzel: Physica C **183**, 99 (1991)

5.55  C. Thomsen: Light scattering in silver halides, in *Light Scattering in Solids VI*, ed. by M. Cardona, G. Güntherodt, Topics Appl. Phys., Vol.68 (Springer, Berlin, Heidelberg 1991) Chap.6

5.56  R. Zeyher, G. Zwicknagel: Solid State Commun. **66**, 617 (1988); Z. Physik B **78**, 175 (1990)

5.57  I.I. Mazin, O. Jepsen, O.K. Andersen, A.I. Liechtenstein, S.N. Rashkeev: Phys. Rev. B **45**, 5103 (1992)

5.58  E. Seider, M. Bauer, L. Genzel, P. Wyder, A. Jansen, C. Richter: Solid State Commun. **72**, 95 (1989); Z. Physik B **83**, 1 (1991)

5.59  T. Zetterer, M. Franz, J. Schützmann, W. Ose, H.H. Otto, K.F. Renk: Phys. Rev. B **41**, 9499 (1990)

5.60  O. Klein, S. Donovan, M. Dressel, G. Grüner: Int'l J. Iinfrared and mm Waves **14**, 2423 (1993)

5.61  D.A. Bonn, R. Liang, T.M. Riseman, D.J. Baar, D.C. Morgan, K. Zhang, P. Dosanjh, T.L. Duty, A. McFarlane, G.D. Morris, J.K. Brewer, W.N. Hardy: Phys. Rev. B **47**, 11314 (1993)

5.62  W.N. Hardy, D.A. Bonn, D.C. Morgan, R. Liang, K. Zhang: Phys. Rev. Lett. **70**, 3999 (1993)

5.63  M.C. Nuss, K.W. Goosen, P.M. Mankewich, M.L. O'Malley: Appl. Phys. Lett. **58**, 2561 (1991)

5.64  A. Röseler: *Infrared Spectroscopic Ellipsometry* (Akademie, Berlin 1990)

5.65  F. Ferrien: Rev. Sci. Instr. **60**, 3212 (1989)

5.66  J. Bremer, O. Hunderi, F. Fanping: Mater. Sci. Eng. B **5**, 285 (1990)

5.67  R. Henn, J. Kircher, M. Cardona, A.M. Gerrits, A. Wittlin, A.V.H.M. Duijn, A.A. Menovsky: Proc. M$^2$HTSC'94. Physica C **235**, 1195-1196 (1994)

5.68  J.H. Kim, B.J. Feenstra H.S. Somal, D. van der Marel, Wen Y. Lee, A.M. Gerrits, A. Wittlin: Phys. Rev. B **49**, 13065 (1994)

5.69  The very good approximation for the superconductor response has been derived by E.H. Brandt (private commun. 1990)

5.70  T. Timusk, D.B. Tanner: In *Infrared Properties of High-T$_c$ Superconductors I*, ed. by D.M. Ginsberg (World Scientific, Singapore, 1989) p.339

5.71  K.F. Renk: In *Studies of High-Temperature Superconductors*, ed. by A.V. Narlikar (Nova Sci., New York 1992) Vol.10

# 6. Magneto-Optical Millimeter-Wave Spectroscopy

Claus Dahl, Philippe Goy, and Jörg P. Kotthaus
With 37 Figures

## List of Symbols

| | |
|---|---|
| $a$ | Lattice period; major axis of an ellipse |
| $\mathbf{A}$ | Vector potential |
| $b$ | Lattice period; minor axis of an ellipse |
| $B$ | Magnetic-field strength |
| $c$ | Speed of light |
| $d$ | Diameter: substrate thickness |
| $\mathbf{E}$ | Electric field |
| $\mathbf{E}_{\text{ext}}$ | External electric field |
| $f_{\text{r}}$ | Receiver frequency |
| $f_{\text{if}}$ | Intermediate frequency |
| $F_0$ | Resonance frequency of Lorentzian oscillator |
| $F_1$ | Frequency of master oscillator |
| $F_2$ | Frequency of slave oscillator |
| $F_{\text{G}}$ | Frequency of Gunn oscillator |
| FIR | Far-infrared |
| FT | Fourier transform |
| $h$ | Lateral depletion length |
| $\mathbf{k}$ | Substrate wavevector |
| $k_0$ | Free space wavevector |
| $\mathbf{k}^*$ | Wavegilde wavevector |
| $l$ | Lacalization length of the edge magnetoplasmon |
| $L_N$ | Conversion loss on harmonic $N$ |
| $\mathbf{L}$ | Depolization tensor |
| $L_{||}$ | Depolization factor in $x$- and $y$-direction |
| $m$ | Harmonic index |
| $m^*$ | Effective mass |
| $n$ | Harmonic index; index of refraction |
| $n(\mathbf{r})$ | Density fluctuation |
| $n_s$ | Two-dimensional carrier density |
| $n_v$ | Three-dimensional carrier density |
| $\mathbf{p}$ | Dipole moment |
| $p$ | Polarization |
| $q$ | Plasmon wavevector |

Topics in Applied Physics, Vol. 74
**Millimeter and Submillimeter Wave Spectroscopy of Solids** Ed.: G. Grüner
© Springer-Verlag Berlin Heidelberg 1998

| $Q$ | Quality factor |
| QHE | Quantum-Hall-effect |
| $r$ | Reflection coefficient |
| $t$ | Transmission coefficient matrix |
| $T_{xx}$ | Diagonal transmission |
| $T_{xy}$ | Hall-transmission |
| $v_F$ | Fermi velocity |
| $\hat{z}$ | Unit vector in $z$-direction |
| $Z_0$ | Free-space impedance |
| $Z^*$ | Waveguide impedance |
| 2DES | Two-dimensional electron system |
| $\Delta\omega$ | Linewidth |
| $\Delta_{xy}$ | Laplace's operator in two dimensions |
| $\varepsilon$ | Dielectric constant |
| $\varepsilon_0$ | Vacuum permittivity |
| $\lambda_F$ | Fermi wavelength |
| $\lambda$ | Wavelength in substrate |
| $\lambda_0$ | Free space wavelength |
| $\mu$ | Mobility |
| $\mu_0$ | Vacuum permeability |
| $\Phi$ | Scalar (electrostatic) potential; phase |
| $\boldsymbol{\sigma}$ | Magneto-conductivity tensor |
| $\boldsymbol{\sigma}^{\text{ext}}$ | Magneto-conductivity tensor with respect to the external field |
| $\sigma_0$ | DC-Drude conductivity at $B = 0$ |
| $\tau$ | Momentum relaxation time |
| $\boldsymbol{\xi}$ | lattice tensor |
| $\omega_c$ | Cyclotron frequency |
| $\omega_{nm}$ | Frequencies of plasma modes with radial and azimuthal index $n$ and $m$ |
| $\omega_p$ | Plasma frequency of the 2D electron gas |
| $\omega_{p,3D}$ | Plasma frequency of the 3D electron gas |
| $\omega_x, \omega_y$ | Fundamental plasma frequencies for polarization in $x$- and $y$-direction |
| $\omega_0$ | Fundamental plasma frequency; frequency of the harmonic oscillator |
| $\omega_\pm$ | Frequency of upper and lower fundamental magnetoplasma mode |
| $\Omega_0$ | Normalizing frequency |

Magneto-optical spectroscopy in solids concerns itself with the study of excitations whose frequency is strongly dependent on the applied magnetic field $B$, and lies in the range of either the cyclotron frequency $\omega_c = eB/m^*$ of a charge carrier with effective mass $m^*$, or of the Larmor frequency $\omega_L = g\mu B$ of a magnetic moment $\mu$. With laboratory magnetic fields nowadays well exceeding $10\,T$ such excitations span the millimeter wave range. For a free electron both the cyclotron frequency and the Larmor frequency are about $280\,GHz$ at a magnetic field of $10\,T$. Conventional spectroscopy in this regime makes use of microwave tubes such as klystrons or backward wave oscillators and, more recently, with the advent of high-frequency solid state sources, of Gunn oscillators or IMPATT diodes all of which have a rather narrow tuning range [6.1, 2]. At higher frequencies, typically above $500\,GHz$ one is forced to use rather clumsy far-infrared lasers with a range of discrete frequencies or has to employ a broad-band thermal source and perform spectroscopy with a Fourier transform spectrometer [6.3, 4] at the expense of limited resolution and relatively small signal to noise ratios. All these techniques are well advanced and succesfully employed, as demonstrated in other chapters of this book but become exceedingly difficult if one wants to do high-resolution linear spectroscopy at arbitrary frequencies in the $100$–$500\,GHz$ range. With a new type of millimeter-submillimeter-wave vector analyzer it has recently become possible to perform high-resolution linear spectroscopy of resonant excitations in solids in the frequency regime, between $30\,GHz$ and about $800\,GHz$, with relative ease. Such a vector analyzer is based on solid state electronics generating high frequencies by multiplication and detecting the frequency band of interest by heterodyne downconversion. Such an instrument bridges the gap between conventional microwave spectroscopy using standard vector analyzers and far-infrared spectroscopy with Fourier transform interferometers. In the following we will discuss the basic techniques and specific applications. We will demonstrate in detail how this type of spectrometer can be effectively employed to study collective electronic excitations in mesoscopic semiconductor structures.

The following section will describe how vector measurements can be achieved at frequencies well above $100\,GHz$. The third section will discuss how the transmission of millimeter wave radiation via both free space and oversized waveguides is achieved and coupling to the samples of interest is optimized. A forth section will focus on some general applications of such a vector instrument ranging from interferometry to measurements of frequency dependent dielectric constants, cavity parameters and losses in superconducting resonators. These applications will demonstrate the flexibility in the use of such a vector instrument as well as its limitations.

In the fifth section we will discuss in detail millimeter- and submillimeter-wave spectroscopic experiments on low-dimensional electron systems as realized in mesoscopic semiconductor structures under the influence of external magnetic fields. We begin with studies of the frequency dependence of the magneto-conductivity tensor of a homogeneous two-dimensional electron system realized in high mobility GaAs/AlGaAs heterostructures. We then focus on the spectroscopy of resonant magnetoplasma excitations in laterally confined two-dimensional electron systems. Such plasma excitations of electronic systems with typical dimensions in the regime of 10 µm have characteristic frequencies in the regime of a few 100 GHz. We will discuss results on basic geometries such as stripes, disks, and rings. These finite size electron systems represent the semi-classical analogues of quantum wires and dots which are studied extensively at higher frequencies with far-infrared spectroscopy [6.5, 6]. We will show that theories based on a local conductivity tensor and a mean field description can adequately describe essential experimental results. The experiments discussed will mainly employ transmission techniques in which the millimeter-wave radiation passes through an array of a large number of identical mesoscopic electron systems in order to achieve sufficient signal strength. We conclude with initial experiments employing near field coupling techniques aimed at studying the millimeter-frequency response of individual mesoscopic electronic devices which are much smaller than the wavelength of the applied radiation.

## 6.1 Vector Analyzer for Millimeter-Wave Frequencies

We present here an instrument with all-solid-state components that can perform phase sensitive measurements at any frequency in the 8–800 GHz interval: the Vector Network Analyzer model MVNA-8-350. It employs two centimeter wave sweepers supplying about 100 mW in the 8–18 GHz interval. They can be used directly, but usually feed Schottky diode multipliers in waveguide mounts with successively smaller waveguide dimensions, so that complete frequency coverage is obtained up to 200 GHz. Measurements at some discrete frequency bands can also be done up to about 350 GHz. For a higher dynamic range in the 200–350 GHz range or experiments up to about 550 GHz, one can alternatively use as a source a millimeter Gunn oscillator associated with the analyzer and feeding a submillimeter wave Schottky diode multiplier. Experiments up to about 800 GHz are possible when using two millimeter Gunn oscillators associated with the analyzer; the 1st as above, the 2nd as a local oscillator of the mixer detector. In contrast to the centimeter wave sweepers, millimeter Gunn oscillators are not broad frequency coverage devices. However, they can be tuned electrically by about ±100 MHz, and mechanically by a few GHz. This opens the possibility of local sweeps, in contrast to the very much fixed frequencies of molecular gas lasers.

The network analyzer is a system including a source, a detector, a frequency drive and a data acquisition and processing system. For a complete information on the changes experienced by the microwave due to any device placed between the source and the detector, the detection must furnish amplitude and phase measurements. Vector measurements have been possible for many years by using interferometric arrangements where a millimeter wave is split into two paths, one of which includes the device under test, and a second fixed one which acts as a reference, with both recombined before detection. The new principle involved in the analyzer discussed here permits to get rid of any dual path configuration in the millimeter wave regime. At detection, the millimeter-wave of interest is downconverted via a Schottky diode harmonic mixer working in a way similar to the frequency multiplier, and sent into a heterodyne vector receiver using an original, very simple, and effective reference channel.

The analyzer is controlled via a personal computer. The software presents many data storage and treatment possibilities such as Fourier Transform (FT) analysis. After a frequency sweep this FT capability permits one to observe time-domain responses, which can reveal itself as being very efficient when testing microwave propagation benches. Complex Lorentzian fitting of resonances (circles in the polar plane) allows all relevant parameters of any resonance to be obtained without ambiguity, for instance a cyclotron resonance or a resonance in a cavity, even when the resonant signal is superposed with a nonresonant contribution.

### 6.1.1 Broadband Sources via Harmonic Generation

For linear spectroscopy, where a large microwave power is not necessary, the use of a low-frequency source followed by a frequency multiplier (multiplication by an integer $N$) to generate frequencies at millimeter wavelengths is very attractive:

$$F_N = N \cdot F_1 . \tag{6.1}$$

Frequency multiplication is accomplished by nonlinear devices. We here use low-capacitance Schottky diodes, consisting essentially of an $n$-doped Si- or GaAs crystal and a tungsten whisker mounted into a waveguide section, which is particularly designed to optimize output power and bandwidth. The output opening corresponds to standard waveguide dimensions and is chosen according to the frequency band of interest. The primary frequency $F_1$ (8–18 GHz) is furnished by a centimeter wave sweeper $S_1$ supplying a power of about 100 mW. This centimeter wave is propagated via a flexible coax cable (which can be as long as 5 m) and coupled into the Schottky diode through an SMA coax connector. The complete multiplier (diode + housing) is also called a "Harmonic Generator" (HG). The waveguide cutoff at lower frequencies ensures that no fundamental frequency $F_1$ can propagate outside the HG. Since the conversion efficiency of the diodes is rather poor, this is important to eliminate spurious signals at the detector side as will be discussed later.

Each standard waveguide millimeter band extends from a frequency $F_a$ to a frequency $F_b$ which is about $1.5 \times F_a$ (Table 6.1). Within this band propagation occurs uniquely in the $TE_{10}$-mode. The limits $F_a$ and $F_b$ are, however, not rigid since the waveguide cutoff frequency is at about $0.8 \times F_a$, and overmoded propagation ($TE_{20}$ and $TE_{01}$) begins above $1.6 \times F_a$.

Since the Schottky diode across the waveguide represents a highly reflective impedance mismatch, a full-band Faraday isolator must be associated to the source in order to reduce standing wave problems. Such devices present a low insertion loss (only a few dB at most), but are sensitive to stray magnetic fields (above a few Gauss). In case of multiplier use very close to a Bitter magnet, for instance, one can replace the isolators by fixed attenuators, with an attenuation of 6 dB or more, realizing a good matching of the HG, but reducing the available dynamic range.

Our apparatus employs a set of multipliers each covering approximately one standard band as listed in Table 6.1. Up to $N = 4$ ($F_N = 72\,GHz$, V-band), the Harmonic Generators are "Flat Broadband" devices, i.e., the output power varies only moderately with the frequency over the full bandwidth. The W-band source can be realized flat broadband as well if one uses cascaded multipliers: a tripler feeding a doubler or a doubler feeding a tripler. Alternatively, one may work with a single diode using harmonics 5 and 6. Frequency sweeps with such a source are, however, limited to about 20 GHz, since for larger intervals and different harmonics the device must be tuned mechanically with a moveable short. The flat broadband multipliers need no tuning adjustment.

Together with the detection scheme to be discussed in the following section, the concept of multiplication by Schottky diodes furnishes a continuous frequency coverage up to $\approx 200\,GHz$, and at particular discrete frequencies detectable millimeter waves are supplied up to $\approx 350\,GHz$. This high-frequency range (110–350 GHz) is realized with a single tuneable multiplier with D-band waveguide output (oversized above 170 GHz) exploiting harmonics 7–20. Unlike the flat broadband sources, whose output is to a good approximation spectrally pure, the tuneable sources emit a comb of millimeter waves. This poses no problem for the linear spectroscopy since the receiver selects the harmonic of interest. In nonlinear experiments (e.g., photo-conductivity) tuneable cavities or band pass filters must be used to obtain monochromatic radiation. Such elements are, however, not commercially available for frequencies above the D-band. Moreover, because of the comparatively low output power of these passive diode sources, one would, in this case, resort to active devices anyway (Gunn-oscillators, carcinotrons, molecular lasers).

## 6.1.2 Heterodyne Vector Detection

Figure 6.1 displays the schematic diagram of the vector analyzer which employs heterodyne detection. To this end, a second Schottky diode ("Harmonic Mixer", HM), practically identical to the multiplier source, is fed by a second centimeter sweeper $S_2$ with frequency $F_2$ (local oscillator). The incoming

**Table 6.1** Frequency coverage of the Schottky diode sources used in conjunction with the vector analyzer

| Frequency band of interest [GHz] | Used mm-Wave harmonic generator | Output waveguide | Corresponding standard millimeter band | Corresponding standard frequency coverage [GHz] | Inner waveguide dimensions [mm] | Used harmonics order $N$ from $S_1$ | Available microwave power at the output [mW] |
|---|---|---|---|---|---|---|---|
| 8–18 | none | SMA coax | - | - | - | 1 | 50 |
| 16–29 | HG-K-FB | WR-42 | K | 18–26.5 | 4.32 × 10.67 | 2 | several |
| 29–51 | HG-Q-FB | WR-22 | Q | 33–50 | 2.84 × 5.69 | 3 | a few |
| 47–72 | HG-V-FB | WR-15 | V | 50–75 | 1.88 × 3.76 | 4 | 0.2–0.8 |
| 70–110 in a single sweep | HG-W-FB | WR-10 | W | 75–110 | 1.27 × 2.54 | 6 | 0.02–0.4 |
| 70–110 in about 3 sweeps | HG-W tunable | WR-10 | W | 75–110 | 1.27 × 2.54 | 4, 5, 6 | 0.002–0.4 |
| 110–170 in about 15 sweeps | HG-D tunable | WR-6 | D | 110–170 | 0.83 × 1.65 | 7–10 | 0.007–0.0002 |
| 170–350 at the best frequency points | HG-D tunable | WR-6 | D | 110–170 | 0.83 × 1.65 | 10–23 | 0.0002–0.000008 |

228    Claus Dahl, Philippe Goy, and Jörg P. Kotthaus

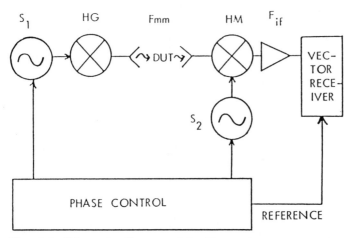

**Fig. 6.1.** Schematic diagram of the Vector Analyzer, in all its possible configurations. The millimeter wave of interest, at the frequency Fmm, is delivered by a Schottky diode Harmonic Generator HG powered by a lower frequency oscillator $S1$, which can be a centimeter sweeper, or a millimeter Gunn diode oscillator. The heterodyne detection front end is a Harmonic Mixer HM, with a local oscillator $S2$ like $S1$. The phase reference of the vector receiver comes directly from the oscillator driving all phase controls

comb of millimeter waves, having traversed the Device Under Test (DUT), then generates frequencies

$$f_{if} = |N F_1 \pm N' F_2| . \tag{6.2}$$

The basic idea of the present analyzer, model MVNA-8-350, AB Millimètre, has been to choose the same harmonic $N' = N$ for multiplication and down-conversion, and to use a pair of identical centimeter sweepers to drive the multiplier and the mixer (French Patent from September, 1989 Centre National de la Recherche Scientifique - Ecole Normale Supérieure; US Patent n°5,119,035, June 2, 1992, also patented in Europe and in Japan). The centimeter sweepers $S_1$ and $S_2$ are phase-locked to each other so that the phase noise in the intermediate frequency

$$\delta\phi_{if} = |N \delta\phi_1 - N \delta\phi_2| \tag{6.3}$$

cancels. The local oscillator $S_2$ is shifted in frequency with respect to $S_1$ such that the mixing product $N |F_1 - F_2|$ corresponds to the fixed frequency $f_r$ of the subsequent vector receiver, where the signal is further downconverted and then digitized for computer processing. The amplitude of the intermediate-frequency signal $f_{if}$ is proportional to the amplitude of the detected millimeter-wave $NF_1$. This proportionality is maintained over several orders of magnitude, if the power of the local oscillator is large with respect to the incoming millimeter wave (MMW) [6.7]. The phase reference comes directly from a 50 MHz reference oscillator controlling the overall phase lock of the system.

The link between the source $S2$ and the HM is realized with the same type of coax cable as the one connecting the HG to the source $S1$. However, it can be interesting, from the point of view of stability, not to use the same length for both cables, especially for experiments with a long distance between HG and HM. The length of the HM cable will be longer than the HG cable in order to compensate for the rapid phase change with frequency due to the distance between HG and HM. Practically an extra HM cable length of 1 m compensates for 1.2 m propagation in a vacuum between the two millimeter heads.

As compared to other millimeter vector network analyzers, the advantages of this configuration are as follows:

- There is no need for any microwave low-noise synthesizer.
- There is no need for any directional coupler and additional harmonic mixer for sampling the emitted millimeter-wave (such devices hardly exist in the submillimeter).
- There is a larger dynamic range, since the use of a narrow-band receiver is possible.
- There is a larger frequency coverage, since many harmonic orders $N$ can be used.
- There are smaller and lighter millimeter heads, linked to the analyzer via flexible coax cables as long as 5 m, which are easy to connect to any place (for instance at the top of a cryostat).

The dynamic range (or signal to noise ratio) depends on the emitted millimeter power $P_N$ and the conversion loss $L_N$ in the mixer at the harmonic $N$, as well as on the noise floor of the heterodyne vector receiver $P_r$,

$$S/N \,[\mathrm{dB}] = P_N \,[\mathrm{dBm}] - L_N \,[\mathrm{dB}] - P_{\mathrm{r}} \,[\mathrm{dBm}] , \qquad (6.4)$$

where $P\,[\mathrm{dBm}] = 10\log P\,[\mathrm{mW}]$. The noise floor of the MVNA-8-350 vector receiver is very low, $P_{\mathrm{r}} = -150\,\mathrm{dBm}$ in its equivalent 10 Hz bandwidth (at an acquisition rate of about ten points per second), yielding a comfortable dynamic range, whose frequency dependence is shown in Fig. 6.2, branches (a) and (b).

Experimentally, the resolution, i.e., the smallest detectable signal *change*, is often more relevant. Besides the dynamic range, it depends on other factors like the mechanical stability of the waveguide setup and the resolution of the subsequent digital signal processing. Further, since the principle of operation relies on a relative stabilization between the two centimeter sources, the absolute stability is merely that of a free running source. To avoid drifting at long measuring times, particularly in the phase signal if the MMW path between HG and HM is long, the absolute frequency may be stabilized by means of an external counter-stabilizer. For measurements at fixed frequency, we thereby attain, for lower frequencies, a resolution of 0.01 dB in amplitude (corresponding to 0.1% change) and 0.1 deg in phase. In measurement with

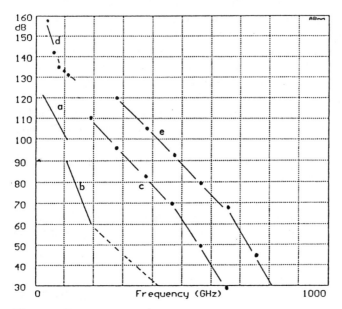

**Fig. 6.2.** Dynamic range (Signal-to-noise ratio) of the Vector Analyzer. Branches $a$ and $b$, Analyzer with Schottky diode generators and mixers. Branch $a$ comprises the bands K, Q, V and W using broadband devices, whereas $b$ corresponds to tuneable D-band millimeter heads. Branches $c$, $d$, and $e$ are given with extensions using Gunn oscillators

variable frequency the repeatability of two sweeps is decisive since the measured spectrum is usually to be compared with a reference spectrum in order to eliminate the frequency response of the diodes and the transmission line. In broadband sweeps (of the order of 10 GHz), standing wave effects and, for the tuneable diodes, the strong variation of the multiplying and mixing efficiency with frequency, considerably reduce the available resolution.

The linear mapping of the MMW spectrum into the RF range relies on the absence of mixing products of higher order in the detector diode. First-order mixing dominates sufficiently if the power of the different harmonics is at least comparable. Since the generated MMW power decreases with increasing harmonic index, this condition is fulfilled if the system is operated not too far above waveguide cutoff. Otherwise, a signal observed at, say, 240 GHz in a D-band setup using $N = 14$, may simply be the 2nd harmonic of the stronger $N = 7$ signal at 120 GHz. To suppress such spurious signals, the lower harmonics must be filtered, e.g., by suitable pieces of narrow waveguide acting as a high-pass filter.

### 6.1.3 Extension to Submillimeter Frequencies and to High Dynamic Range

According to (6.4) the dynamic range improves with increasing power of the source and/or with reduced conversion loss at the mixer. This is achieved by replacing the combination centimeter sweeper/Schottky multiplier by an active millimeter source. The master-oscillator $S_1$ is then used as a local oscillator coupling the active source to the system with an additional phase lock loop. Curve (d) in Fig. 6.2 shows the dynamic range with various Gunn sources and detection with the conventional diodes. The drawback of such a setup is, of course, the loss of wide tunability.

If the Gunn oscillator with frequency $F_G$ feeds itself a multiplier diode, the generated comb of frequencies $M F_G$ may extend to the submillimeter regime. Stabilizing the Gunn oscillator on the harmonic $N$ of $S_1$ with a frequency shift $F_D$, the relevant mixing product using conventional detection is

$$f_{if} = M F_G - M N F_2$$
$$= M [N (F_1 - F_2) - F_D] . \qquad (6.5)$$

As before, $f_{if}$ must be tuned to the receiver frequency $f_r$ by suitably adjusting the frequency difference $F_1 - F_2$ of the centimeter sweepers. In the experiments discussed below, we use a 30 mW Gunn oscillator working at 95 GHz with mechanical tuning $\pm 1$ GHz and electronical tuning $\pm 100$ MHz, locked onto harmonic $N = 6$. The offset $F_D$ is 50 MHz and the receiver frequency $f_r = 9.0105$ MHz. Detection is effected by the tuneable D-band mixer at harmonic $N \times M$. The output power of the submm diode is 8, 0, $-3$, $-10$, $-24$, $-38$, $-57$ dBm, for $M = 2$–8, respectively. Harmonics up to $M = 7$ (at 665 GHz) can be detected with the HM-$D$. Reasonably large signals are restricted to $M < 6$. The observed dynamic range is shown in curve (c) of Fig. 6.2. Naturally, one can choose a Gunn with a frequency different from 95 GHz (below 140 GHz, which is a practical limit for these sources), to match any desired submillimeter frequency.

Associating a second Gunn oscillator together with a submillimeter diode at the detector side further enhances the dynamic range [branch (e) in Fig. 6.2]. Compared to the harmonic mixer HM-$D$ used in the previous configuration, this solution presents a conversion loss $L_{N \times M}$ reduced by about 20–30 dB, improving by the same quantity the dynamic range S/N. For instance at 570 GHz, (harmonic $M = 6$ of a 95 GHz Gunn) one has about S/N = 80 dB, instead of S/N = 50 dB with the previous setup using a single Gunn at the source. The losses $L_{N \times M}$ are about: 29, 39, 42, 45, 48, 48 dB for $M = 3$–8, respectively. Measurements up to 760 GHz are thus possible.

The submillimeter power delivered by the multiplier can be controlled for any $M$ by placing a calibrated variable millimeter attenuator behind the Gunn oscillator. The observed output/input power dependence slope is, after desaturation, close to the multiplication parameter $M$. Figure 6.3 shows the output power in arbitrary units as a function of the attenuation at the

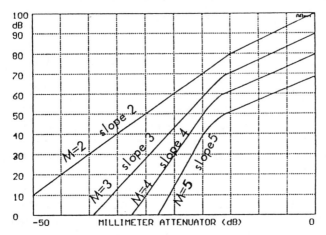

**Fig. 6.3.** Control of the power emitted on different harmonics ($M = 2, 3, 4, 5$) from a submillimeter wave Schottky diode powered by a 95 GHz Gunn oscillator. *Vertically*: detected signal above noise, in dB. *Horizontally*: position, in dB, of the attenuator inserted between the Gunn and the diode

multiplier input. Similarly, a centimeter attenuator between $S_1$ and HG may serve to vary the emitted millimeter power in the conventional configuration of the analyzer. In the same manner, a pin-diode modulator can be employed as a switch if the millimeter or submillimeter wave is to be applied only at a given time.

## 6.2 Coupling the Wave to the Sample

### 6.2.1 The Sample Environment

The microwave response of a solid state sample can be obtained by transmission, if the sample is reasonably transparent, or by reflection, if it is conductive. The sample can be placed across a waveguide (Sect. 6.2.2), in free space (Sect. 6.2.3) or in a cavity. The use of oversized waveguides, rectangular or cylindrical ("light pipes"), will reveal itself to be very interesting, since they are easier to machine and minimize the propagation losses as compared to fundamental-mode waveguides. In order to reduce the mismatches and to maintain the polarization, one must use progressive pyramidal transitions ("tapered transitions") between the small rectangular guide of the harmonic generator or mixer and the oversized waveguide. Free space propagation is also possible using feed horns and quasi optical-lenses or mirrors.

There are many advantages in using cavities. They permit one to control the configuration of the electromagnetic field and to increase the sensitivity by multiplying the number of crossings through a transparent sample or, in the case of a conductive sample which plays the role of one of the cavity walls, by increasing the effective field experienced by the sample. In the last case the sample is observed by reflection.

The drawback in cavity use is mostly the restriction in working frequencies, since one needs a careful frequency tuning on resonant modes. From this point of view, Fabry-Pérot cavities equipped with a movable mirror offer the advantage of an easy, continuous and wide tunability. When the coupling into the cavity is operated via coupling holes of a given diameter $D$ optimized for a given wavelength (for instance $D = 0.3\lambda$), the theoretical very wide-band coverage of the Fabry-Pérot is practically restricted to about one octave.

There might also be another drawback in using cavities, which is easy to take into account. Let us consider an experiment at a fixed frequency, at which a tuned cavity containing a sample is submitted to a magnetic field. Depending on the cavity filling factor, the changes of sample properties with the magnetic field at resonance may perturb the cavity tuning. In order to unambiguously measure the aborption-dispersion parameters (and not a mixture of them), one must lock the analyzer frequency onto the cavity frequency. Then the detected amplitude purely represents the absorption (the detected phase remaining constant, due to the lock), and the correction voltage for maintaining the tuning gives the dipersion across the sample resonance.

Any type of cavity, rectangular, cylindrical, Fabry-Pérot (preferably close to semi-confocal or confocal geometries) can be observed by reflection (there is, for instance, a single coupling hole), or by transmission (with two coupling holes).

In the following Sect. 6.2.2, the words transmission-reflection apply to the propagation system from the analyzer millimeter heads, HG and HM, to the sample environment: waveguide, free space, or cavity.

### 6.2.2 Propagation via Waveguides

It is possible to observe, by sharp minimum reflections, resonances in a cavity coupled to a single waveguide. However, care must be taken to distinguish, for instance, by a mechanical change in cavity dimensions, the cavity mode from other possible standing-wave effects.

If the propagation is performed in a normal guided mode, one can use a directional coupler of the millimeter band of interest in order to obtain the reflected signal. In case the propagation is performed in an oversized waveguide (or light pipe), especially in the submillimeter regime, one can use a 45° mylar sheet as a diplexer. The source can be put in the horizontal section of the waveguide. After the 45° semi-transparent mirror, an absorber (like the black polymer used for integrated circuit grounding) closes the 2nd horizontal branch. The lower vertical branch goes to the cavity, and the upper vertical branch contains the detector. This setup is very easy to place along the vertical axis of a cryostat containing a single waveguide (or light pipe) terminated by the cavity at the cryostat bottom.

The most favorite setup for solid state spectroscopy with magnetic fields consists of a transmission system using waveguides. It can also be used with cavities pierced with two coupling holes. Then the cavity resonances are observed as transmission maxima.

We first give several examples of the simplest possible waveguide transmission setup, working without cavity. This setup can be used at many different frequencies without any careful adjustment. If possible, the waveguide size is chosen large enough to keep the frequency cutoff well below the lowest frequency to be used. A good compromise with usual room restrictions is the Q-band waveguide (Table 6.1) used from 27 to 480 GHz at Ludwig-Maximilians-Universität, Munich and also at Clarendon Laboratory, Oxford. Another possible choice, with more space, is the K-band waveguide, used from 16 to 300 GHz at the High Magnetic Fields Laboratory, Nijmegen. (Also used there are brass-made light pipes of 13 mm diameter, with similar results). Tapered transitions (half cone angle about 3°) must be used between the millimeter head waveguide and the propagation waveguide (oversized most of the time). The latter can have standard bends or 45° mirrors. As in any propagation system, care must be taken to minimize standing waves.

Before its use for spectroscopy experiments, the analyzer must be employed for characterizing and optimizing the setups. Let us first consider the setup used in Munich, made from about 1.25 m Q-band waveguide to go down to the superconductive coil where the wave crosses the sample, then a severe 180° bend (8 mm inner curvature radius), then 1.25 m Q-band waveguide coming back up to the top. It is possible to obtain a clear inside view of this sample holder.

Thanks to a directional coupler attached onto the waveguide input at the top of the cryostat, the reflected microwave is detected through a frequency sweep such as 34–38 GHz (Fig. 6.4, upper curve). In the Fourier transform of this signal one observes reflection peaks versus time, and can thus deduce the places of propagation mismatch (lower curve).

Reflections are observed in (a) and (b), corresponding to mismatches in the waveguide made from brass and stainless steel pieces, and mostly in (c) where there is the bend at the cryostat bottom. Mismatches (d) and (e) correspond to the symmetrical waveguide sections as (b) and (a). The peak at (f) corresponds to residual reflection from the end of the waveguide which is terminated with a matched load. This sample holder also presents losses by transmisson (not shown): 5 dB around 35 GHz, 8 dB around 96 GHz, 25 dB around 170 GHz and 35 dB at 475 Ghz.

Figure 6.5, curve (a), presents a transmission measurement performed at Oxford on a 18 mm outer diameter sample holder about 1.65 m high, made from two parallel Q-band stainless steel wave-guides with thin walls (about 0.4 mm) connected via two 45° mirrors at the cryostat bottom where the superconductive coil is located.

Curve (b) in Fig. 6.5 has been obtained by placing in series with the waveguide a rectangular cavity made from a 6 mm long piece of silver plated

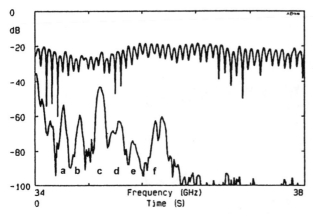

**Fig. 6.4.** Reflection study of a 1.25 m high, U-shaped, sample holder at Munich, fabricated from pieces of Q-band (WR-22) copper and stainless steel waveguides. At *top* is shown the reflected signal, in dB, against frequency, with the other side of the sample holder terminated by a matched load. At *bottom* is shown, downshifted for clarity, the Fourier transform of this signal, with a horizontal time scale. This horizontal scale corresponds to a distance scale (2 m back and forth per division, i.e., 1 m along the sample holder), taking into account the wave speed inside the waveguide. Positions of the propagation mismatches, *a–f*, can be clearly identified

WR-10 waveguide closed with thin walls pierced with 1 mm diameter coupling holes. The peaks $M1$ at 54.3 GHz (maximum electrical field at the cavity center) and $M2$ at 69 GHz (maximum magnetic field) are the two lowest frequency modes of this cavity which presents a 20 dB transmission loss at resonance. (This 20 dB value is a typical good compromise between a lower value which could be obtained with larger coupling holes, then decreasing the

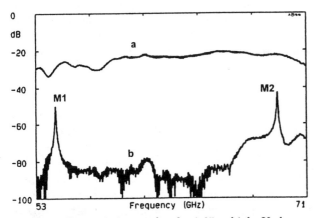

**Fig. 6.5.** Transmission study of a 1.65 m high, U-shape, sample holder at Oxford, made from Q-band (WR-22) stainless steel waveguides. *a* signal transmitted during a 53–71 GHz frequency sweep. *b* same measurement with a rectangular cavity inserted across the waveguide. $M1$ and $M2$ are the first two cavity resonance modes

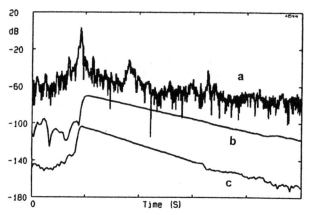

**Fig. 6.6.** Fourier Transforms (FT) of the transmitted signals shown in Fig. 6.5. *Curve a* is the FT of Fig. 6.5a, and shows a peak at the time necessary to cross the sample holder. *Curves b* and *c*, downshifted for clarity, are the FT of the signal around $M1$ and $M2$ in Fig. 6.5b, respectively

quality factor $Q$, and a higher value with smaller coupling holes, improving $Q$, but making the experiments more difficult).

For measuring its conductivity versus magnetic field, a small, highly conductive sample contained in a capillary quartz pipe will be introduced into the cavity center. The mode $M1$ (maximum electric field at the cavity center) could be shifted by the quartz dielectric constant. However, the sample conductivity dampens the resonance so that $M1$ disappears. The mode $M2$ (maximum magnetic field) is practically not moved in frequency and permits one to study the sample conductivity.

Figure 6.6 shows in (a) the FT of Fig. 6.5a (no cavity), and in (b) and (c) the FT of $M1$ and $M2$ from Fig. 6.5b (empty cavity in series). The peak in (a) indicates the time necessary to cross the sample holder. After this time, the decay time constants $\tau$ (necessary time for $1/e$ decay in power, i.e., 4.34 dB drop) are 4.4 ns for (b) and 3.2 ns for (c). They give the resonance quality factors $Q$ at the center frequency $F_0$ as

$$Q = 2\pi F_0 \tau. \tag{6.6}$$

One obtains $Q = 1500$ and $1400$ for $M1$ and $M2$, respectively (empty cavity).

To perform a transmission experiment it is desirable that the sample covers a reasonable fraction of the waveguide cross section. At lower frequencies, where the waveguide size increases, a larger specimen is thus needed to maintain experimental sensitivity. Further, in high-magnetic-field experiments the available bore in the superconducting coil is limited to 40 mm at most in diameter. Practical waveguide sizes for the magnetospectroscopy of solids therefore usually do not extend much below the K-band, so that for frequencies below the corresponding cutoff, one must resort to coaxial cables. Then, the coupling of the microwave to the sample can be realized by means of microstriplines connected to the cables by commercially available transitions. We employ a triplate stripline, also called screened stripline [6.8], where the center conductor is located symmetrically between two ground planes. The width $w_s$ of the stripline relative to its distance $d_s$ to the ground planes is chosen so as to obtain an impedance of $50\,\Omega$. The absolute dimensions are kept rather small ($w_s = 1.3\,\mathrm{mm}$, $d_s = 0.8\,\mathrm{mm}$) in order to avoid the excitation of higher modes at the frequencies of interest. The sample is placed into a slit perpendicular to the stripline. Alternatively, the stripline may be directly fabricated onto the sample. Examples of experiments employing both config- urations will be presented in Sect. 6.4. The described microstripline setups present considerable insertion losses since the transmission is based on capac- itive coupling between the ends of the interrupted center conductor due to stray fields. Moreover, the stripline discontinuity presents a severe mismatch, causing a strongly frequency-dependent response of the whole transmission system. Nevertheless, magnetospectroscopy experiments in the stripline con- figuration have been possible in a wide frequency range, extending up to 70 GHz.

### 6.2.3 Propagation in Free Space

Millimeter, and more so submillimeter waves, are a domain of the electro- magnetic spectrum where quasi-optical techniques can be extremely useful. Good scalar (corrugated) feed horns are commonly available up to 200 GHz. In a relatively large frequency interval (+/-20%), each scalar horn produces a Gaussian beam which permits one to couple waveguide propagation into free space. In the submillimeter regime, Potter (dual mode) horns are easier to fabricate and give very low sidelobes in a narrow frequency band (a few percents). This situation is exactly adapted to our needs, since our submil- limeter frequencies are mostly a comb of harmonics of the associated Gunn frequencies, for instance $95 \pm 1$ GHz.

When refocusing the divergent Gaussian beam by means of mirrors or lenses (made from low-loss dielectrics like teflon or high-density polyethylen, HDPE), free space propagation can be performed with lower losses than through guided modes in waveguides. For cryogenics reasons, it can also be very interesting to couple the microwave beam into a sample without the need of a conducting metal like the waveguide. However, to reduce diffrac- tion problems, the free space propagation needs more space than the guided propagation. Moreover, the optical windows, which are necessary for a quasi- optical coupling onto a sample placed in vacuum (for instance in a cryostat), can create problems of standing waves and thermal losses by radiation.

We have used free space propagation setups with two or four focusing lenses. One of the advantages of these setups is that the beam waist at the sample can be very small (below 3 mm at 475 GHz). For transmission studies through samples of dimensions still smaller than the beam waist, one can place a diaphragm around the sample. Attenuation, observed at 91, 380, and

475 GHz through diaphragms of diameter $D$, placed at mid-distance between two identical HDPE lenses, is 10, 20 and 30 dB for $D = 1.5\lambda$, $0.7\lambda$ and $0.4\lambda$, respectively. Here, the lenses possess a focal length of 100 mm and a diameter of 78 mm, and are placed 400 mm apart. The source horn is at 200 mm from the first, and the detector horn at 200 mm from the second lens.

A second setup employs four corrugated (corrugations calculated to reduce reflections at the center of the band 75–110 GHz) HDPE lenses $L1$-$L2$-$L3$-$L4$ of 78 mm diameter and 100 mm focal length. The source horn aperture is at the focus of $L1$, the detector horn aperture at the focus of $L4$ and the distance between $L2$ and $L3$ is 200 mm. This system can be adapted to cryogenic needs (windows, etc.), since the distances $L1$-$L2$ and $L3$-$L4$, where the beam is parallel, can be chosen at will. The insertion loss source/detection, including the scalar horns and the four lenses, is of the order of 1.8 dB as compared to the direct contact source/detection. In order to be able to work at any frequency, we have also used non-corrugated lenses in the same geometry.

Transmission calibration is made simply when recording the instrumental response in the absence of the sample. Then one records the effect due to the introduction of the sample. A practical suppression of parasitic standing wave effects is obtained by digital filtering of the FT of the signal. In the FT of the recorded signal, one must cut the contributions corresponding to times longer than the crossing of the sample, with eventual multiple reflections in it. Back-transformation then yields the filtered signal.

To investigate reflection in free space one can use a single lens, with the horn aperture on one side and the sample placed symmetrically at the other side, at twice the focal length of the lens. The calibration can be performed simply by comparing the reflection from the sample to the reflection from a (supposedly perfect) mirror.

## 6.3 General Applications

The phase measurement possible with a vector analyzer contains a distance information, since any change $x$ in the source/detector distance gives a corresponding change in phase of the detected vector signal $S$,

$$S = A \exp\left[\mathrm{i}\left(\omega t - kx\right)\right], \tag{6.7}$$

where $k = 2\pi/\lambda$ is the wave vector. As shown in the following sections, it is on resonant signals that the phase information demonstrates all its possibilities.

### 6.3.1 Interferometry

A signal at 285 GHz is transmitted through a Michelson interferometer and recorded versus time, i.e., versus the position of the movable mirror. Different representations of the same signal can help one understand what can be concluded from phase measurement. Figure 6.7a shows the amplitude in linear units [amplitude $A$ of (6.7)] and the phase in degrees in a Cartesian plot. Figure 6.7b shows the same data as a polar plot, where the distance from the center is the linear amplitude $A$ of (6.7), and the angle is the detected phase, $360°x/\lambda$, expressed in degrees. The parameter on the curve is the time, or the mirror position. The physical understanding is clear, since one "sees" the vector sum of a fixed vector $OO'$ (signal due to the branch with fixed mirror), and the movable vector $O'M$ (signal due to the movable mirror).

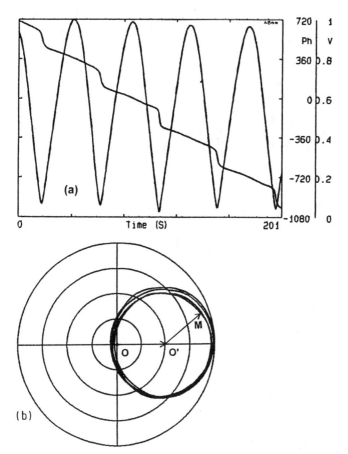

**Fig. 6.7 a,b.** Linear amplitude and phase of the transmission at 285 GHz through a Michelson interferometer when linearly moving one of the two mirrors. In the Cartesian plot (**a**) the amplitude exhibits a $\lambda_0/2$ periodicity while the phase decreases monotonically. In the polar plot (**b**) a $\lambda_0 2$-move of the mirror results in a complete circle

### 6.3.2 Dielectrics

Figure 6.8 shows the reflected and the transmitted microwave signal of a
$d = 9.97\,\text{mm}$ thick slab of sapphire, a low-loss dielectric material with a large
permittivity $\varepsilon = 9.40$. This slab is a Fabry-Pérot interferometer, with an
optical distance between front and back mirrors (the air/sapphire planes)
of $d\sqrt{\varepsilon} = 30.56\,\text{mm}$. The variations in the reflected signal observed here
versus frequency, displayed as Cartesian (Fig. 6.8a, bottom curve) and as
polar plot (Fig. 6.8b, circular curve), are very similar to the ones observed in
the transmitted amplitude vs. position through the Michelson interferometer,
Fig. 6.7. The reflection is a periodic function of the microwave frequency with
period $\Delta F$:

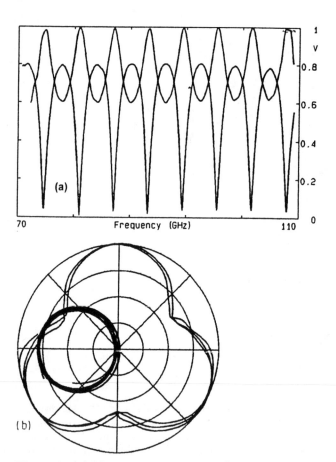

**Fig. 6.8.** (a) Microwave transmission (*top curve*) and reflection (*bottom curve*)
amplitudes through a 9.97 mm thick sapphire sample at normal incidence, varying
the microwave frequency from 70 to 110 GHz. (b) Polar plot of (a), where the
reflected signal describes a circle and the transmitted signal a cycloid

$$\Delta F = c/\left(2d\sqrt{\varepsilon}\right) , \tag{6.8}$$

where $c/\sqrt{\varepsilon}$ is the propagation velocity in the dielectric of refractive index $n = \sqrt{\varepsilon}$. Maximum amplitude reflection coefficient $r$ is observed at frequencies which are an odd number of $\Delta F/2$, with a value

$$r = (\varepsilon - 1)/(\varepsilon + 1) . \tag{6.9}$$

Equation (6.9) gives $r = 0.808$ which is observed in Fig. 6.8a.

The transmission through this sapphire slab shows a combination of resonant effects with a linear effect: the replacement of air by sapphire with the thickness $d$ involves an observable change in phase $\Delta\phi$ (in degrees) such that:

$$\left(\sqrt{\varepsilon} - 1\right) d/\lambda = (\Delta\phi/360°) + m , \tag{6.10}$$

where $m$ is an integer number of turns. From the phase variation with frequency one can obtain the value of the integer $m$. Then the most precise value of $\varepsilon$ can be deduced from (6.10), when taking the phase value at the frequencies where changes due to oscillations cancel, i.e., at the maxima and minima in transmission (Fig. 6.8a, top curve). The period of transmission oscillations $\Delta F$ is given by (6.8). Transmission maxima are observed at reflection minima, i.e., at integer values of $\Delta F$. Of course, reflection and transmission are coupled, $t^2 + r^2 = 1$, which is practically observed in Fig. 6.8a.

The above interference effects must be taken into account when designing a microwave window. A good transmission at a fixed frequency is easy when using a low-loss dielectric with a thickness $d$ such that transmission is maximum. For broadband transmission experiments one must prefer dielectrics with lower permittivities $\varepsilon$, such as 2.06 in Teflon (0.56 dB interference ripple), 2.24 in high density polyethylen (0.7 dB), 3.80 in fused quartz (1.8 dB).

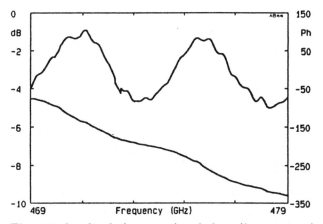

**Fig. 6.9.** Amplitude (*top curve*) and phase (*bottom curve*) of the normal-incidence transmission through the sample of Fig. 6.8 at submillimeter frequencies, where, contrary to Fig. 6.8, the material is seen to exhibit measurable damping

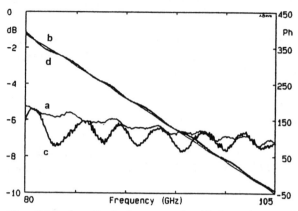

**Fig. 6.10.** Amplitude (*curves a,c*) and phase (*curves b,d*) of the normal-incidence transmission through a 21 mm thick Araldite slab. A first measurement is performed with a large size (about 15 mm) microwave beam (*a,b*). A second measurement (*c,d*; more wavy results) is performed with a 3 mm diameter metallic diaphragm, demonstrating the possibility to characterize samples smaller than the wavelength

As we can see in (6.10), permittivity measurements are greatly simplified with phase information. Naturally, the amplitude information can give the loss in the sample, as soon as it is noticeable. Figure 6.9 shows a transmission measurement through the sapphire sample of Fig. 6.8, realized at high frequency, so that the attenuation $\alpha$ is no longer negligible and can be measured at the transmission maxima. One obtains the loss $\tan \delta = 0.0008$ in this dielectric at this frequency from

$$\tan \delta = 1.1\alpha \, [\mathrm{dB/cm}]/\left(F \, [\mathrm{GHz}] \cdot \sqrt{\varepsilon}\right) \, . \tag{6.11}$$

Loss measurements in rather lossy materials are simpler than in very low-loss materials, since the attenuation is larger, and since the standing waves inside the sample are damped. Figure 6.10 shows, with a 20.95 mm thick araldite lossy sample, transmission amplitude (a) and phase (b), from which we deduce the permittivity $\varepsilon = 2.90$ and the loss $\tan \delta = 0.02$. Interference effects are hardly visible in transmission.

Curves a and b in Fig. 6.10 were obtained from a large size (diameter: 50 mm) sample, about 15 mm diameter of which was illuminated by the microwave at the common focus of the lenses. Good transmission measurements are also possible with very small samples (below $\lambda$ in size). Curves c and d in Fig. 6.10 show the transmission through the same araldite sample, with a 3 mm diameter aluminum diaphragm pressed against it. One can see that the diaphragm increases the standing waves inside the sample, and shifts their phase. However, it does not modify the overall phase variation, giving a correct measurement of $\varepsilon$ via (6.10), and shows similar amplitude maxima, from which the dielectric losses are correctly obtained via (6.11).

### 6.3.3 Cavities and Superconductors

As discussed in Sect. 6.2.1, cavities are extremely useful in spectroscopy, since they allow the interaction between the electromagnetic waves and the specimen to be enhanced (for instance low-loss dielectrics can be characterized in open cavity experiments). As for other resonances, vector measurements can comprehensively characterize cavity resonances. Consider the $TE_{121}$ mode in a brass-made cylindrical cavity of about 7 mm, both in diameter and in height, with two coupling holes pierced onto the axis. Figure 6.11a shows amplitude and phase of the observed transmission resonance. The theoretical fit of amplitude $A$ and phase $\phi$ to a Lorentzian

$$A = 1 \Big/ \sqrt{1 + 4Q^2 \left[ (F - F_0)/F_0 \right]^2},$$
$$\Phi = -\arctan \left[ 2Q \left( F - F_0 \right)/F_0 \right],$$

(6.12)

yields, besides the resonance frequency $F_0$, a quality factor of $Q = 4400$. The fit cannot be distinguished from the experimental trace. The polar plot of this resonance (Fig. 6.11b) exhibits a perfect circle. The origin $O$ belongs to the circle, and corresponds to infinite or zero-frequency extrapolations, with zero transmission.

With about the same geometry as in Fig. 6.11, a niobium-made cylindrical cavity, cooled at 4.2 K, would show the same resonance profile, with an extremely narrow width, because of the high quality factor $Q = 2360000$. However, the observed amplitude and phase, displayed in Fig. 6.12a, are distorted as compared to the preceding resonance. Such signals originate from the superposition of the resonant wave with a nonresonant leak wave bypassing the cavity. Leakage here is of the same order of magnitude as the resonant signal because the cavity is extremely weakly coupled in order to avoid cavity damping. In a purely scalar measurement the distorted lineshape renders the accurate determination of the resonance parameters rather difficult. Our measurement, being vector, shows a perfect fit is possible, superimposed to amplitude and phase in Fig. 6.12a, and also in the polar plot, Fig. 6.12b. In the polar representation, one can immediately see that the leak wave amplitude $OL$ is comparable to the resonance amplitude (diameter of the circle). The resonance frequency (given by the fitting software) is at $R$ (opposite to $L$), and not at the maximum amplitude $M$. Fitting problems arise if the leak wave signal is not constant over the swept bandwidth. The above vector-fitting technique to suppress leakage applies to any type of resonance (versus frequency, magnetic field, or any other parameter), provided its lineshape, not necessarily Lorentzian, is known, thus solving one of the problems frequently encountered in solid state spectroscopy.

In the following example, we show the capability of the vector fit to distinguish between mixed resonances. A transmission measurement around 475 GHz through a Fabry-Pérot resonator permits one to observe two modes, the smaller one being distorted by the vicinity of the larger, see Fig. 6.13,

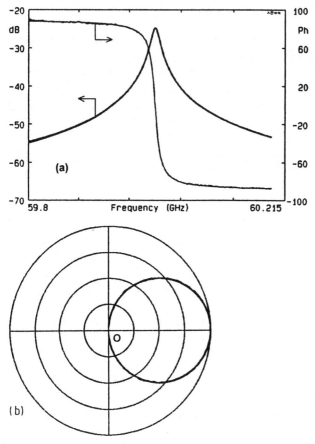

**Fig. 6.11 a,b.** Amplitude and phase variations when observing a cavity resonance by transmission (TE$_{121}$ mode in a brass-made cylindrical cavity, with two coupling holes). (**a**) Cartesian, (**b**) polar plot

where the experimental signal appears in (a), recorded amplitude, and in (b), recorded phase. The good fits in amplitude (c) and in phase (d) are obtained as follows. The vector fit of the large resonance is shown in Fig. 6.14 in amplitude (a) and phase (b). One moves this phase (b) by 180°, and makes the vector difference of the experimental curve and this fit. On the resulting curve one makes the fit of the small resonance, with amplitude (c) and phase (d) in Fig. 6.14. Then the global fit, as a control, is obtained by the vector sum of the two fits, amplitude (c) and phase (d) in Fig. 6.13. This procedure yields all relevant parameters of the two resonances, their resonance frequencies, their quality factors ($Q = 18,000$ for the large resonance, $Q = 36,000$ for the small one), and their relative amplitudes. (The small resonance is 17 dB below the large one in Fig. 6.14, to be compared to the small peak in Fig. 6.13 which is only 12.5 dB below the large one due to the mixing of modes).

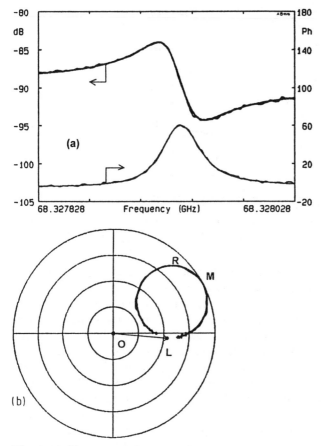

**Fig. 6.12.** Experimental trace and Lorentzian fit of the amplitude and phase transmitted through a superconducting cavity around resonance, (**a**) Cartesian, (**b**) polar plot. The detected wave is a combination of resonant and non-resonant contributions ($TE_{121}$ mode in a niobium-made cylindrical cavity, with two extremely small coupling holes). In the polar plot the vector $OL$ describes the nonresonant contribution. Contrary to Fig. 6.11 the resonance position $R$ is no longer identical with the maximum amplitude $M$

### 6.3.4 Bulk Semiconductors in High Magnetic Fields

Our analyzer is extremely well-adapted to microwave spectroscopy with high magnetic fields. The frequency coverage of the analyzer well supplements far-infrared Fourier transform techniques, which become difficult at small wavenumbers. Any frequency can be matched, and sweeps versus frequency are possible (contrary to, e.g., laser sources). The equipment is not critically sensitive to stray magnetic fields and can work at a reasonably close distance from big magnets. The vector measurement possibilities permit very quick,

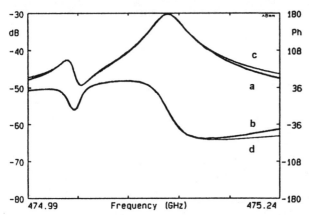

**Fig. 6.13.** Split resonance observed by transmission through a Fabry-Pérot resonator, *a* amplitude, *b* phase, with corresponding *c*, *d*, double Lorentzian fit

unambiguous and complete understanding of the observed phenomena as demonstrated in the remaining part of the article.

At the Nijmegen High-Field Magnet Laboratory the microwave transmission was studied through a highly doped semiconductor sample of InSb at room temperature, placed in a 0–20 T sweepable magnetic field [6.9]. In Fig. 6.15a the transmitted amplitude through the highly degenerate InSb slab is shown at frequencies of 35 GHz and 81 GHz, respectively. With an increasing magnetic field the sample becomes more transparent and exhibits more or less strong oscillations in the transmitted signal. At first sight this behavior could be ascribed to some optical absorption between states which are split in the magnetic field. An analysis of the phase of the transmitted signal in a polar plot (Fig. 6.15b) shows, however, that the oscillatory transmission

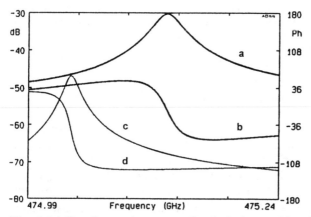

**Fig. 6.14.** Two Lorentzians: *a* amplitude, *b* phase, with a $Q = 18\,000$ quality factor, *c* amplitude, *d* phase, with $Q = 36\,000$. Their vector sum is the fit in Fig. 6.13

signal results from an interference effect. The strong dependence of the real part of the dielectric constant of the conduction-electron plasma on magnetic field causes strongly field-dependent multiple reflections in the sample slab. This phenomenon bears the name "helicon waves" as becomes immediately obvious from the polar plot shown.

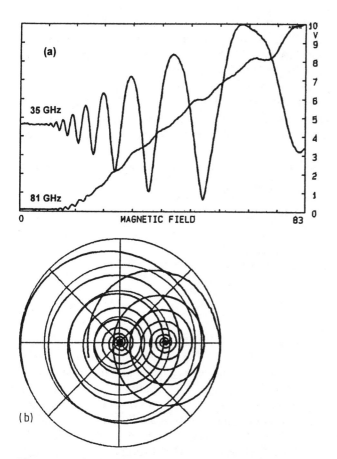

**Fig. 6.15 a,b.** InSb sample at room temperature submitted to a 0–17 T magnetic field sweep at fixed frequencies. (**a**) Transmitted microwave amplitude at 35 GHz (strongly oscillatory behavior) and at 81 GHz (monotonic behavior with only small oscillations). (**b**) Corresponding polar plots 81 GHz at left, 35 GHz at right. Zero magnetic field corresponds to the origin of the spirals

# 6.4 Magnetospectroscopy
# of Low-Dimensional Electron Systems

## 6.4.1 Magnetoconductivity
## of the Two-Dimensional Electron System

Two-dimensional electron systems (2DES's) have been realized at various semiconductor surfaces or hetero-interfaces. The Si-MOS structure and the modulation-doped GaAs/GaAlAs heterostructure constitute the systems that have been most intensively investigated [6.10]. In this section we briefly review some high-frequency properties of the homogeneous 2DES [6.4, 6, 11]. Laterally structured systems will be considered in Sects. 6.4.2, 3. Throughout this last part of the chapter on magneto-optical millimeter wave spectroscopy all experimental results shown have been obtained for 2DES's in GaAs/GaAlAs heterostructures.

Within the Drude model the local dynamic magneto-conductivity tensor of the 2DES is given by

$$\sigma_{xx} = \sigma_{yy} = \sigma_0 \frac{1 - i\omega\tau}{(1 - i\omega\tau)^2 + \omega_c^2\tau^2}$$

$$\sigma_{xy} = -\sigma_{yx} = -\sigma_0 \frac{\omega_c\tau}{(1 - i\omega\tau)^2 + \omega_c^2\tau^2} \tag{6.13}$$

where

$$\sigma_0 = \frac{n_s e^2 \tau}{m^*} \tag{6.14}$$

and $n_s$, $m^*$ and $\tau$ are the surface carrier density, effective mass and momentum relaxation time, respectively. For $\omega\tau \gg 1$ resonance absorption occurs at the cyclotron frequency $\omega_c = eB/m^*$, i.e., $\mathrm{Re}\,\sigma_{xx}$ has a maximum at $\omega = \omega_c$ with a linewidth (FWHM) of $\Delta\omega = 2/\tau$. In a 2DES, Landau quantization can become visible in the cyclotron-resonance signal in a sweep of the magnetic field at moderate $\omega\tau$, in that $1/B$ periodic quantum oscillations modulate the cyclotron resonance amplitude [6.12]. For $\omega\tau < 1$ these transform into the well-known Shubnikov de Haas oscillations of the dc-magneto-conductivity. These quantum oscillations can be measured with relative ease at microwave frequencies [6.13] and are often used to determine the surface carrier density $n_s$ in a microwave experiment.

a) **Relation Between Conductivity and Transmission, Cyclotron Resonance.** The frequency dependence of the transmission of the whole experimental setup can exceed the transmission change originating from the 2DES by several orders of magnitude. It is usually eliminated by normalizing the spectra or working at fixed frequency and varying a parameter which acts only on the 2DES. Here we mostly show data with a varying magnetic field $B$, but in MOS- or HEMT-structures one may alternatively employ $n_s$-sweeps by varying the applied gate voltage. Even so, a rigorous connection between conductivity and transmission is difficult to obtain because the standing wave pattern due to multiple reflections is itself influenced by the conductivity of the electron gas. In the following this is demonstrated for the simple case of multiple reflections in the substrate in the quasi-optical regime, which can be treated analytically. We consider a 2D conducting layer of conductivity $\sigma$ (e.g., as given by 6.13) on top of a dielectric substrate with thickness $d$ and index of refraction $n$. The half-spaces below and above the substrate are assumed to have refractive indices $n'$ and 1, respectively. We neglect the small distance ($< 100\,\text{nm}$) between the 2DES and the upper surface of the substrate, as well as absorption other than from the 2DES. From the continuity conditions for the components of the electromagnetic field the relation between transmitted and incident field for normal incidence is obtained as

$$\boldsymbol{E}^{\text{t}} = M^{-1}\boldsymbol{E}^{\text{i}} = \mathbf{t}\boldsymbol{E}^{\text{i}}, \tag{6.15}$$

where $\mathbf{t}$ is the transmission matrix, given as the inverse of the matrix $\mathbf{M}$ which has the components

$$
\begin{aligned}
M_{xx} &= \frac{1}{2}e^{ik'd}\left[(1 + n' + Z_0\sigma_{xx})\cos kd \right. \\
&\quad \left. -i\frac{n'}{n}\left(\frac{n^2}{n'} + 1 + Z_0\sigma_{xx}\right)\sin kd\right], \\
M_{yy} &= \frac{1}{2}e^{ik'd}\left[(1 + n' + Z_0\sigma_{yy})\cos kd \right. \\
&\quad \left. -i\frac{n'}{n}\left(\frac{n^2}{n'} + 1 + Z_0\sigma_{yy}\right)\sin kd\right], \\
M_{xy} &= -M_{yx} = \frac{1}{2}e^{ik'd}\left(Z_0\sigma_{xy}\cos kd - i\frac{n'}{n}Z_0\sigma_{xy}\sin kd\right).
\end{aligned}
\tag{6.16}
$$

In (6.16), $k = k_0n = n\omega/c$ and $k' = k_0n' = n'\omega/c$ are the wavevectors in the substrate and the lower half-space, respectively, and $Z_0 = (\mu_0/\varepsilon_0)^{1/2} = 377\,\Omega$ is the vacuum impedance. The usual experimental situation is $n' = 1$. For $n' = n$, on the other hand, we obtain the case without internal reflections where (6.15, 16) simplify to give [6.14]

$$\boldsymbol{E}^{\text{t}} = 2\left(1 + n + Z_0\sigma\right)^{-1}\boldsymbol{E}^{\text{i}}. \tag{6.17}$$

Experimentally, the Fabry-Perot interferences in the substrate can be suppressed by wedging it at an angle of $2$–$3°$. Multiple reflections from other discontinuities in the waveguide setup can, however, lead to similar interference effects. Equations (6.15–17) show that $\sigma_{xx}$ and $\sigma_{xy}$ simultaneously enter the expressions for both diagonal and non-diagonal transmission. For high magnetic fields such that $|Z_0\sigma_{xy}| \ll |Z_0\sigma_{xx}| \ll 1$ and $|Z_0\sigma_{xy}| \ll |Z_0\sigma_{yy}| \ll 1$, (6.17) can be approximated by

$$t_{xx} = \frac{2}{1-n}\left(1 - \frac{1}{1+n}Z_0\sigma_{xx}\right),$$

$$t_{yy} = \frac{2}{1-n}\left(1 - \frac{1}{1+n}Z_0\sigma_{yy}\right), \tag{6.18}$$

$$t_{xy} = \frac{2}{(1+n)^2}Z_0\sigma_{xy}\left[1 - \frac{1}{1+n}Z_0(\sigma_{xx} + \sigma_{yy})\right],$$

so that in this case the change in the diagonal transmission is given by $\sigma_{xx}$ or $\sigma_{yy}$ and the Hall transmission is dominated by $\sigma_{xy}$.

A rigorous treatment of the problem of magneto-transmission (and -reflection) of the 2DEG in the waveguide configuration has not been given. Complicating factors are the different waveguide wavelengths and impedances for the two polarizations, inside and outside of the substrate, and the possible excitation of higher modes. Even if the waveguide is operated in the fundamental mode (as often will not be the case in the experiments to be discussed later), higher modes may still be excited in the substrate. Some special cases have been treated in the literature. *Meisls* and *Kuchar* [6.15] consider the reflection and transmission of the fundamental mode neglecting internal substrate reflections. *Shanjia* et al. [6.16] perform multimode calculations for more complicated sample geometries but do not include a magnetic field. Assuming pure $TE_{10}$-propagation, (6.15) is still valid if $k$ and $Z_0$ are simply replaced by the wavevector of the filled waveguide, $k^* = \sqrt{(n\omega/c)^2 - (\pi/L)^2}$ and the impedance of the empty waveguide $Z^* = 1/\sqrt{\varepsilon_0/\mu_0 - (\pi/\mu_0\omega L)^2}$, respectively, where $L$ denotes the wide side of the waveguide. This constitutes a reasonable approximation for the case of small signals such that the polarization-mixing terms are negligible, i.e., $Z^*|\sigma_{xy}| \ll 1$.

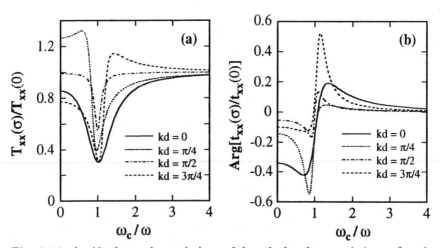

**Fig. 6.16 a,b.** Absolute value and phase of the calculated transmission as function of magnetic field for a 2DEG on a dielectric substrate of different thickness $d$ in the vicinity of cyclotron resonance. Parameters used are $n_s = 3 \times 10^{11}$ cm$^{-2}$, $m^* = 0.07\,m_e$, $\omega\tau = 20$ and $n = 3.53$

The transmitted MMW is a periodic function of the substrate thickness $d$ with period $\lambda/2$ where $\lambda = \lambda_0/n$ is the wavelength in the substrate. Figure 6.16 shows the normalized diagonal magneto-transmission $T_{xx}(\sigma)/T_{xx}(\sigma = 0) = |t_{xx}(\sigma)/t_{xx}(\sigma = 0)|^2$ and the corresponding phase for a 2DES with typical parameters at different substrate thicknesses using (6.15, 16) with $n' = 1$. For $d = 0$ and $d = \lambda/4$ the transmission lineshape is symmetric and purely absorptive, but other values of $d$ lead to dispersive admixtures. Likewise, the phase can acquire a partially absorptive lineshape or may exhibit an asymmetric peak. The resonance frequency and the linewidth of the electron system are thus to be determined [6.12] from a fit of the measured transmission to (6.15). As already discussed in Sect. 6.3.3, the problem of determining the resonance parameters in the presence of lineshape distortions can be simplified considerably if the phase is experimentally available as well. We recall that, when displayed in the polar plot, Lorentzian resonances are represented as circular curves. Here, for cyclotron resonance (CR), the curve parameter is the magnetic field. An asymmetric lineshape only results in a different orientation of the circles with respect to the origin. Figure 6.17 shows how the resonance parameters – position and linewidth – may easily be constructed. The resonance positions thus obtained are independent of $d$ and correspond well to the maxima in $\mathrm{Re}\,\sigma_{xx}$ (for $\omega\tau \gg 1$). Due to saturation effects, the widths may depend on $d$ as can already be seen directly from Fig. 6.16 for $d = 0$ and $d = \lambda/4$. More generally, resonances of the electron

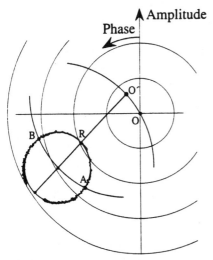

**Fig. 6.17.** Measured cyclotron resonance of a GaAs/GaAlAs-heterostructure at 98 GHz displayed in a polar plot (linear amplitude vs phase; curve parameter is the magnetic field). Also shown is the construction of resonance position (magnetic field at $R$) and linewidth (difference between magnetic fields at $A$ and $B$). The resonance position does not correspond to the minimum of the amplitude. i.e.. amplitude and phase would have an asymmetric shape (from [6.17])

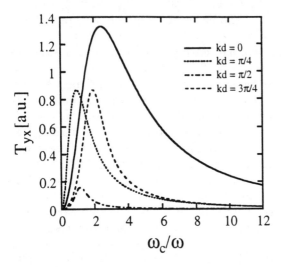

**Fig. 6.18.** Calculated Hall transmission $T_{yx}$ of a 2DEG on a dielectric substrate. Parameters are $n_s = 5 \times 10^{11}\,\mathrm{cm}^{-2}$, $\omega\tau = 5$, $\omega/2\pi = 30\,\mathrm{GHz}$ and $m^* = 0.067\,m_e$

system having Lorentzian shape may thus be quantitatively analyzed in the presence of multiple reflections of unknown origin.

**b) Hall Conductivity.** As for the diagonal transmission $T_{xx}$, the lineshape of the Hall transmission $T_{yx} = T_{xy} = |t_{xy}|^2$ is also strongly influenced by Fabry-Perot resonances in the substrate as shown in Fig. 6.18. Only for magnetic field regimes where $|Z_0\sigma_{xy}| \ll 1$ and where $\sigma_{xx}$ does not strongly vary with $B$, i.e., in the vicinity of $B = 0$ and for high magnetic fields ($\omega_c \ll, \gg \omega, 1/\tau$), the non-diagonal transmission is directly given in terms of $\sigma_{xy}$, $T_{yx} \propto |\sigma_{xy}|^2$. Figure 6.19 shows the experimental Hall transmission of a GaAs/GaAlAs heterostructure at $f = 33\,\mathrm{GHz}$ [6.17]. To measure $T_{yx}$ the rectangular waveguide is operated in the $\mathrm{TE}_{10}$-mode, and the waveguide sections before and behind the sample are crossed with respect to each other.

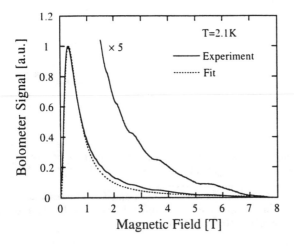

**Fig. 6.19.** Experimental Hall transmission of a heterostructure together with a fit according to (6.15) using the Drude conductivity tensor. Parameters are discussed in the text. The quantum-Hall-plateau visible at highest magnetic field corresponds to a Landau level filling factor 4

A frequency-dependent residual background signal, e.g., due to nonperfect waveguide crossing, is minimized by selecting a suitable frequency. The measured signal corresponds well to the theoretical curves of Fig. 6.18 for the parameters $kd = 0$ and $\omega\tau = 0.45$. Above $B = 1\text{T}$ the signal exhibits plateaus due to the Quantum Hall Effect (QHE). From their $1/B$-periodicity one obtains the carrier density, here, $n_s = 5.0 \times 10^{11}\,\text{cm}^{-2}$. Similar MMW investigations of the QHE have been carried out by *Kuchar* et al. [6.18]. *Galchenkov* et al. [6.19] measure $T_{yx}$ between 30 and 60 GHz, and observe a decreasing plateau width with increasing frequency. Theoretically, the frequency dependence of the quantum Hall plateaus has been treated by *Vieweger* and *Efetov* [6.20].

### 6.4.2 Magnetoplasma Resonances in a Laterally Confined Electron System

When the 2DES is excited at finite wavevectors, the cyclotron resonance excitation transforms into another type of collective oscillations of the electron gas, the so-called magnetoplasma resonances. At zero magnetic field the dispersion of these density oscillations depends on the plasmon wavevector roughly as $q^{1/2}$ [6.21], contrary to the three-dimensional plasma frequency, which is approximately independent of $q$. Two-dimensional plasmons have first been observed on the surface of liquid helium [6.22] and shortly afterwards in MOS devices [6.11, 23]. A nonzero wavevector $q$ can be introduced either by spatially modulating the external exciting electromagnetic field, e.g., using grating couplers, or by spatially modulating the electron gas itself. In this part, we will study in detail the plasma excitations in the laterally confined 2DES of varying geometry subject to a homogeneous external high-frequency field. The size of the lateral confinement roughly (but not rigorously) defines a plasmon wavelength which here will be of the order of $10\,\mu\text{m}$, so that the plasma frequencies for a typical GaAs/GaAlAs heterostructure are in the 100 GHz regime. On these length scales complex geometries with smallest features of about $1\,\mu\text{m}$ can easily be realized by simple optical lithography, even in large arrays of identical objects. These are usually needed in order to obtain sufficient signal strength in millimeter wave transmission spectroscopy. Investigations of arrays of submicrometer structures have so far mostly been restricted to simple, namely linear or circular geometries. As will be discussed below, pattern transfer from a mask with submicrometer features to two-dimensional electron systems in heterostructures is severely limited by the lateral depletion of mobile electrons which usually occurs on length scales of several 100 nm. The possibility of millimeter wave spectroscopy of single mesoscopic objects will be addressed in Sect. 6.4.3.

The system dimensions in question are, by far, larger than the Fermi wavelength of the electron gas so that the plasma excitations can be described within a classical hydrodynamic model [6.24–27]. In this approximation the electron gas is treated as a charged fluid, characterized by a total density

$n_s + n(r, t)$, and a velocity field $v(r, t)$. The ground state is given by the 2D density $n_s$ and a vanishing mean velocity, i.e., $n = 0$ and $v = 0$. The equilibrium density $n_s$ is assumed to be spatially homogeneous in the interior of the system, and compensated by the positive background charge (ionized donors). The dynamics are described by the equation of motion, applied to each fluid element, the wave equation, and the continuity equation. We consider the linearized equations in $n$ and $v$ which, upon introducing the electrodynamic potentials $A$ and $\Phi$ in the Lorentz gauge, read

$$
\left( \frac{\partial}{\partial t} + \frac{1}{\tau} \right) v = \frac{e}{m^*} \left( \nabla_{xy} \Phi + \frac{\partial A}{\partial t} \right)_{z=0} - \omega_c \, v \times \hat{z} - \frac{s^2}{n_s} \nabla_{xy} n,
$$

$$
\left( \Delta - \frac{\varepsilon}{c^2} \frac{\partial^2}{\partial t^2} \right) \left( \begin{array}{c} \Phi \\ A \end{array} \right) = e\delta(z) \left( \begin{array}{c} n/\varepsilon\varepsilon_0 \\ \mu_0 n_s v \end{array} \right), \tag{6.19}
$$

$$
-\frac{\partial}{\partial t} n = n_s \nabla_{xy} \cdot v.
$$

In (6.19) $\varepsilon$ denotes the background dielectric constant and $s^2 = \frac{3}{4} v_F^2$ takes into account the compressibility of the 2DES [6.21]. Scattering processes are included by means of a phenomenological relaxation time $\tau$. The electron gas is assumed to be located in the plane $z = 0$, subject to a perpendicular, external static and homogeneous magnetic field, which is treated separately from the vector potential $A$. The equation of motion and the continuity equation are restricted to two dimensions, whereas the wave equation must be solved in whole space.

In the regime $\omega\sqrt{\varepsilon}/c \ll q \ll \omega/v_F$, which is relevant for the systems considered here, the effects of retardation and compressibility are negligible so that (6.19) can be treated here in the limit $c \to \infty$, $s \to 0$. As in Sect. 6.4.1 we assume a time dependence $\sim e^{-i\omega t}$ of all quantities. The continuity equation thus allows $v$ to be eliminated to obtain the "plasma equation" and the Poisson equation coupling $n$ and $\Phi$:

$$
\frac{n_s e}{m^*} \Delta_{xy} \Phi = \left( \omega^2 + \frac{i\omega}{\tau} - \frac{\omega_c^2}{1 + i/\omega\tau} \right) n,
$$

$$
\Delta \Phi = \frac{e n \delta(z)}{\varepsilon\varepsilon_0}. \tag{6.20}
$$

For an infinite 2DEG, where a spatial dependence $\sim e^{iqx}$ can be assumed, and for $\omega\tau \gg 1$, this yields for the magnetoplasma excitations at wavevector $q$

$$
\omega^2 = \omega_p^2 + \omega_c^2, \tag{6.21}
$$

where

$$
\omega_p^2 = \frac{n_s e^2}{2\varepsilon\varepsilon_0 m^*} q \tag{6.22}
$$

is the 2D plasma frequency at $B = 0$.

For confined systems (6.20) has been treated analytically only for the case of a half plane [6.28, 29]. Numerical solutions have been obtained for circular [6.30], ring [6.31, 32], and strip geometries [6.33], using "hard wall" boundary conditions. The problem of the correct boundary conditions that must be imposed in the case of confined systems is still under theoretical discussion [6.34]. Also, the assumption of a position-independent equilibrium density can be a poor approximation, as will be seen.

a) **Circular Disks.** Some qualitative features of magnetoplasma resonances in finite 2D systems can be obtained with a particularly simple model system, namely a 3D ellipsoid of constant volume carrier density $n_v$. Since, in a homogeneous external field, an ellipsoid is homogeneously polarized, a mode exists with homogeneous electric field $\boldsymbol{E} = -L\boldsymbol{P}/\varepsilon\varepsilon_0$ and thus homogeneous carrier motion. Here, $L$ is the depolarization tensor and $\boldsymbol{P} = -n_v e\boldsymbol{x}$ the polarization where $\boldsymbol{x} = i\boldsymbol{v}/\omega$ denotes the displacement of the carriers from their equilibrium position. The depolarization tensor describes the influence of the geometry on the depolarization field and hence the restoring force connected with the carrier displacement with respect to the background charge. The equation of motion in this "ellipsoidal model",

$$\left(\frac{\partial}{\partial t} + \frac{1}{\tau}\right)\boldsymbol{v} = -\frac{e}{m^*}\left(\boldsymbol{E}_{\text{ext}} - L\boldsymbol{P}/\varepsilon\varepsilon_0\right) - \omega_c\boldsymbol{v} \times \hat{\boldsymbol{z}}, \tag{6.23}$$

establishes a relationship between the internal current and the external field, $\boldsymbol{j} = \sigma^{\text{ext}}\boldsymbol{E}_{\text{ext}}$, which defines the conductivity of the system. We first consider the case of a thin oblate spheroid, where the two lateral depolarization factors coincide, $L_x = L_y =: L_{\parallel}$; less symmetric structures will be considered later. This model system approximates, to some extent, a circular 2D disk. Its conductivity is readily obtained as

$$\sigma^{\text{ext}} = \frac{i\omega\sigma_0/\tau}{\left(\omega^2 - \omega_0^2 + i\omega/\tau\right)^2 - \omega^2\omega_c^2}$$

$$\begin{pmatrix} \omega^2 - \omega_0^2 + i\omega/\tau & -i\omega\omega_c \\ i\omega\omega_c & \omega^2 - \omega_0^2 + i\omega/\tau \end{pmatrix}. \tag{6.24}$$

Note that with $\omega_0 = 0$, one recovers the Drude conductivity of the homogeneous electron gas, (6.13). Alternatively, we may consider the complex solutions of (6.23) for $\boldsymbol{E}_{\text{ext}} = 0$. They are obtained as

$$\omega_{\pm} = \pm\tfrac{1}{2}\left(\omega_c \mp i/\tau\right) + \sqrt{\left(\omega_c \mp i/\tau\right)^2/4 + \omega_0^2}. \tag{6.25}$$

yielding the eigenfrequencies of the plasma oscillation in the 2D-plane as $\text{Re}\,\omega$, and the damping as $-2\,\text{Im}\,\omega$. For $\omega_0\tau \gg 1$ these quantities correspond to the maxima and widths in $\text{Re}\,\sigma_{xx}$, as given by (6.24). In (6.24, 25), $\sigma_0$ is the *dc* Drude conductivity at $B = 0$ (6.14) and

$$\omega_0^2 = \omega_{p,3D}^2 \cdot L_{\parallel}, \tag{6.26}$$

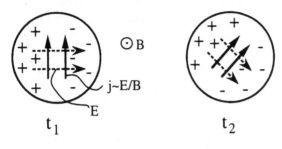

**Fig. 6.20.** Induced depolarization fields, polarization charges, and currents of the magnetoplasma oscillation $\omega_-$ for $\omega_c \gg \omega_0$ at two successive times

the plasma frequency of the spheroid at $B = 0$, where $\omega_{p,3D}^2 = n_v e^2/m^* \varepsilon \varepsilon_0$ is the 3D plasma frequency. For a thin oblate spheroid with diameter $d$ and thickness $2w$ such that $d \gg w$ one has $L_\parallel = \pi w/2d$. Introducing the sheet density $n_s$ via $n_v V = n_s A$ where $V$ is the volume and $A$ the projected area of the ellipsoid, the plasma frequency (6.26) can be rewritten as

$$\omega_0^2 = \frac{3\pi}{8} \frac{n_s e^2}{m^* \varepsilon \varepsilon_0 d}. \tag{6.27}$$

For $B \neq 0$ the plasma resonance splits into a mode with positive and one with negative magnetic field dispersion. The splitting is a consequence of the distinction between right- and left-handed electron motion in the presence of a magnetic field. The two modes are qualitatively different at large $B$. Whereas the upper branch approaches cyclotron resonance, independent of the geometry, the lower branch, $\omega_- \approx \omega_0^2/\omega_c = n_v e L_\parallel/\varepsilon \varepsilon_0 B$, decreases as $1/(Bd)$, independent of the effective mass. The lower mode is driven by the Hall current perpendicular to the depolarization field which leads to a redistribution of the polarization charges and, consequently, to a rotation with the frequency $\omega_-$ of all quantities $(E, v, n)$ that characterize the plasma oscillation as sketched in Fig. 6.20. The electron paths themselves, i.e., the solutions $x(t)$ to (6.23) are right- and left-handed circular orbits with the frequency $\omega_+$ and $\omega_-$, respectively. All carriers move in a coherent fashion and are only slightly displaced from their equilibrium position. Only the resulting polarization charges perform orbits around the edge of the disk. Equation (6.25) is valid independent of the system size, and in the undamped limit equally applies for quantum dots, provided the bare confining potential is harmonic, $V(r) = \frac{1}{2}m^* \omega_0^2 r^2$ (generalized Kohn theorem [6.35–37]). The effects of lateral quantization of the electronic motion are, therefore, better demonstrated in transport experiments [6.38, 39].

Perpendicular to the disk plane, the above model would predict an oscillation with the 3D plasma frequency, since $L_z \approx 1$. Here, the classical description fails because the electrons are confined to a length of the order of the Fermi wavelength. Moreover, the approximatively triangular shape of the vertical confining potential of a heterojunction does not allow for an application of the generalized Kohn theorem. The collective transitions between the ground state and the excited states of the electron gas in a heterojunction- or quantum-well-potential (intersubband plasmons) must be treated quantum-mechanically. Their energies lie in the far- to mid-infrared and have intensively been investigated both experimentally and theoretically [6.10].

The frequencies obtained in (6.25) represent the fundamental resonance of the spheroid. Higher modes (with inhomogeneous internal fields) have been obtained by solving Laplace's equation with appropriate boundary conditions, describing the electron gas in the ellipsoid with the frequency-dependent dielectric function [6.40, 41]. Homogeneous external fields excite only the lowest mode, however.

From the conductivity (6.24) one obtains for the relative linewidths [6.42]

$$\frac{\Delta\omega_+}{\Delta\omega_-} = \frac{\omega_+}{\omega_-} \tag{6.28}$$

and for the oscillator strengths

$$\frac{f_+}{f_-} = \frac{\omega_+}{\omega_-}. \tag{6.29}$$

The resonance amplitudes are the same for both branches. The absolute oscillator strengths follow from the sum rule $f_+ + f_- = \pi n_s e^2/2m^*$. The absolute widths at large $B$ ($\omega_c \gg \omega_0$) are

$$\Delta\omega = \begin{cases} 2/\tau & \text{upper branch,} \\ 2\omega_0^2/\omega_c^2\tau & \text{lower branch,} \end{cases} \tag{6.30}$$

and $\Delta\omega = 1/\tau$ for $B = 0$. Whereas the upper branch acquires the CR linewidth as expected, the lower mode exhibits a narrowing such that there exists a well-defined resonance even for $\omega\tau \ll 1$. Physically, this is a consequence of the fact that for high $B$ the Hall angle approaches 90° so that the Joule dissipation $\boldsymbol{j}\cdot\boldsymbol{E}$ tends to zero. Figure 6.21 shows the real and imaginary part of $\sigma_{xx}^{\text{ext}}$ as function of frequency.

The resonance frequency and linewidth can also be expressed using the local conductivity tensor of the homogeneous electron gas. For the lower mode at high $B$ this yields

$$\omega = -\frac{3\pi}{8}\frac{\sigma_{xy}(\omega = 0)}{\varepsilon\varepsilon_0 d}, \quad \frac{\Delta\omega}{\omega} = -\frac{2\sigma_{xx}(\omega = 0)}{\sigma_{xy}(\omega = 0)}, \tag{6.31}$$

independent of the particular model for $\sigma$ [6.43]. It should, however, be kept in mind that (6.23–31) are a particular consequence of the assumption of an ellipsoidal density profile. Other models for the ground state electron density may lead to considerably different resonance parameters. In the 2D-theory by Volkov and Mikhailov to be discussed below [6.28, 29], the resonance frequency of the lower mode is modified by a $B$-dependent logarithmic correction, and its linewidth is obtained as $\Delta\omega \sim \sigma_{xy}/\varepsilon\varepsilon_0 d$, i.e., independent of the diagonal conductivity $\sigma_{xx}$. Quantum theories of magnetoplasmons in laterally confined systems (with parabolic and non-parabolic confinement [6.44]) have mostly excluded the question of damping.

Magnetoplasma modes in the confined 2DES were first observed by *Allen* et al. [6.45] in 1983 on 3 μm-disks at far-infrared (FIR) frequencies. They have

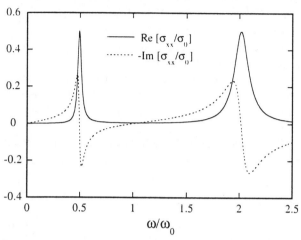

**Fig. 6.21.** Real and imaginary part of the complex dynamic conductivity of a thin ellipsoid, (6.24) for $\omega_c/\omega_0 = 1.5$ and $\omega_0\tau = 10$

since then been studied intensively in quantum-dot systems where the size is of the order of the Fermi wavelength as well as for large electron disks with diameters up to 10 mm. The electron systems have been realized on various semiconductor hetero-interfaces (GaAs/GaAlAs, Si/SiO$_2$) or surfaces (InSb), and also on the surface of liquid helium [6.46] (for a review on the physics of quantum dots see [6.47]). We demonstrate here, that with its large bandwidth of several decades in the frequency regime of interest, the combination of FIR with MMW spectroscopy allows for a particularly comprehensive investigation of the magnetoplasmons in such systems, i.e., a measurement over a wide magnetic field range.

In Fig. 6.22 we show amplitude and phase of the MMW transmission at fixed frequency as a function of the magnetic field for circular disks with diameter $d = 11.5\,\mu$m, in a square array having a periodicity of 14 µm. The measurements are performed in the Q-band waveguide setup described in Sect. 6.2.2. The signal at 23 GHz, a frequency which is below waveguide cutoff, has been obtained by the stripline technique also discussed in Sect. 6.2.2. All data are taken at temperature $T = 4.2$ K unless stated otherwise. The disk matrix covers an area of $2.5 \times 4.0$ mm$^2$ and is fabricated on a GaAs/GaAlAs-heterostructure sample whose size slightly exceeds the waveguide cross section. The pattern is defined by conventional optical contact lithography followed by wet chemical etching. Amplitude and phase signal exhibit two resonance branches with positive and negative magnetic field dispersion, respectively. It is interesting to note that the relative phase change is opposite for the two modes, since, in a magnetic field sweep, a resonance signal begins with the high- or low-frequency side, depending on whether the branch has positive or negative magneto-dispersion. In the polar plot, Fig. 6.23, this phase behavior translates into an opposite sense of rotation along the curves in the direc-

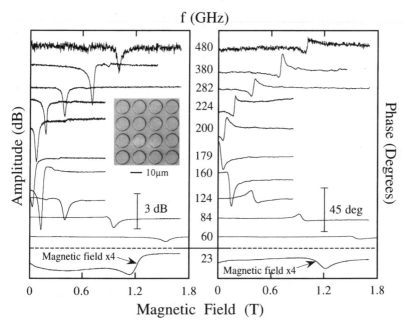

**Fig. 6.22.** Amplitude and relative phase of the transmitted millimeter wave at fixed frequency $f$ as a function of the magnetic field for an array of 11.5 μm circular disks at $T = 4.2$ K. Measurements are performed in a waveguide setup except for the curve at 23 GHz, which is taken in stripline geometry. The *inset* shows a micrograph of the sample surface

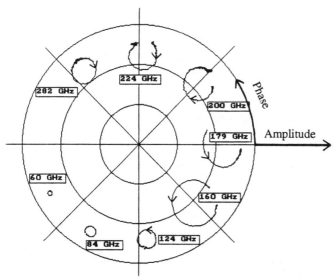

**Fig. 6.23.** Polar plot of the resonances of Fig. 6.22. The sense of rotation along an increasing magnetic field is opposite for $\omega < \omega_0$ and $\omega > \omega_0$, indicating negative and positive magneto-dispersion, respectively

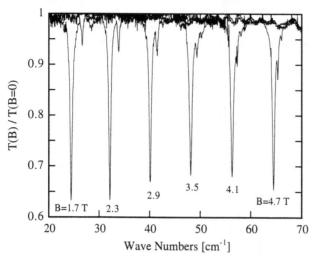

**Fig. 6.24.** Transmission, normalized to $B = 0$, of the same sample as in Fig. 6.22 at FIR frequencies as function of the wave number $1/\lambda$ ($10\,\mathrm{cm}^{-1} \cong 300\,\mathrm{GHz}$). The signal is obtained by Fourier-transform spectroscopy at a resolution of $\Delta(1/\lambda) = 0.1\,\mathrm{cm}^{-1}$ and exhibits three plasma modes of the disks

tion of increasing $B$, which is clockwise (counterclockwise) for the mode with positive (negative) magnetic field dispersion. Clearly visible is the decreasing oscillator strength of the lower branch with increasing $B$. FIR-data measured with Fourier-transform spectroscopy are displayed in Fig. 6.24. This powerful method has largely contributed to the understanding of excitations in lower dimensional electron systems [6.6]. Details on the experimental setup can be found elsewhere [6.4]. The FIR spectra show the further behavior of the upper branch and reveal additional modes which approach the fundamental one at large $B$. All modes with positive magneto-dispersion asymptotically merge into CR. The ensemble of the measured resonance positions is displayed in Fig. 6.25.

The simple model which has lead to (6.25) correctly predicts the appearance of the two fundamental magnetoplasma resonances but does not account for the observation of the higher modes. The more realistic theory of *Fetter* [6.30] assumes a 2D disk with constant sheet density $n_s$ and a step-like density profile and hard wall boundary conditions at the edge. The calculation yields a complete set of eigenfrequencies, $\omega_{n,\pm m}(B)$, $n = 1, 2, \ldots$, $m = 0, 1, \ldots$, where $n$ and $m$ are the radial and azimuthal mode indices, respectively. The magneto-dispersion of some branches, together with the spatial distribution of the induced density for selected values of $B$, are shown in Fig. 6.26. The fundamental $B = 0$ resonance frequency obtained from this theory

$$\omega_{11} = 1.857\Omega_0, \quad \Omega_0^2 = \frac{n_s e^2}{\pi m^* \varepsilon \varepsilon_0 d}, \tag{6.32}$$

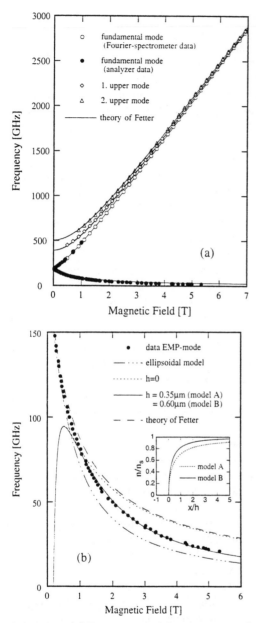

**Fig. 6.25.** (a) Frequencies of the resonances observed in Figs. 6.22–24 in comparison with the model of *Fetter* [6.30] using the effective mass $m^* = 0.068 \, m_e$ and the position of the fundamental mode $\omega_0/2\pi = 185 \, \text{GHz}$ as fit parameters. (b) Details of the lower branch of (a) in comparison with the theories of EMPs discussed in the text. At high magnetic field the best fit to the data is obtained via (6.34, 36) assuming a finite lateral depletion length $h$, whose quantitative determination depends on the model used to describe the edge electrostatics. The inset shows calculated density profiles for the two models discussed in the text. Also shown are the dispersion obtained numerically with the theory of Fetter, whose high-$B$ side coincides with (6.36), and the dispersion for an ellipsoidal density profile

agrees well with the plasma frequency $\omega_0$ of the ellipsoidal model, (6.27). The experimentally observed frequency of $\omega_0 = 185\,\text{GHz}$ is somewhat lower than the theoretical value, $204\,\text{GHz}$, using the appropriate sample parameters. The discrepancy can be explained by interdisk interaction, as will be discussed in Sect. 6.4.2d. The higher radial modes are calculated as $\omega_{12}/\omega_{11} = 2.12$ und $\omega_{13}/\omega_{11} = 2.73$. The calculation of the corresponding modes of an ellipsoid yields considerably higher frequencies [6.40]. Using the fundamental plasma frequency $\omega_{11}$ as a fit parameter, the experimental spectrum (Fig. 6.25) is quite well reproduced by the 2D theory. Only modes having nonzero dipole moment are observed in the waveguide setup where the external high-frequency field is uniform over the size of a disk. This selection rule, $|m| = 1$ for circular geometries, is relaxed for inhomogeneous external fields (Sect. 6.4.3). Further, it is seen in Fig. 6.26 that all modes $\omega_{n,|m|}$ are split into doublets $\omega_{n,\pm m}$ at finite $B$, which is a consequence of the distinction between left- and right-handed electron motion in the presence of a magnetic field. Experimentally, the splitting is not resolved here for the higher branches $n \geq 2$, but has been observed for electron disks on the surface of liquid helium, where the resonances possess sufficiently narrow linewidths. Electron rings, a closely related geometry, exhibit larger splitting, so that the effect can in this case also be observed for semiconductor plasmas, as will be shown in Sect. 6.4.2b. The lower mode, $\omega_{1,-1}(B)$, is not well described by the above theories, neither by the 2D model of Fetter nor by the ellipsoidal model (Fig. 6.25b). To correctly reproduce the experimental points at high $B$ a refined model of the density profile near the edge is needed. Vice versa, the precise measurement of the magnetoplasma dispersion yields information on parameters concerning the boundary of the 2D electron system, in particular the lateral depletion length, as will be elaborated in the following.

In Fig. 6.26b we remark that, contrary to the CR-like modes, the density fluctuation connected with the modes having negative magneto-dispersion is localized at the edge. These low-frequency modes have, therefore, been termed edge-magnetoplasmons (EMPs) and are sensitive to the details of the density profile near the edge of the confined electron system. It is clear that neither the ellipsoidal nor the abrupt density profile correctly describes the prevailing edge electrostatics. Realistic profiles have been calculated by *Chklovskii* et al. ([6.48], Eq.10), in the following referred to as "model $A$" and *Gelfand* and *Halperin* ([6.49], Eq.7), "model $B$" and are of the form $n(x) = n_\text{s} \times f(x/h)$ where f is a function that rises from 0 to 1 essentially in an interval $x/h \approx 1$ and where the scaling length h is the lateral depletion length. Such profiles influence the EMP frequency when the EMP localization length becomes smaller than $h$. The theory of EMPs as given by *Volkov* and *Mikhailov* [6.28, 29] allows for an analytical description of this effect. They obtain the EMP localization as the absolute value of the complex quantity

$$\ell = \frac{i\sigma_{xx}(\omega)}{2\varepsilon\varepsilon_0\omega}. \tag{6.33}$$

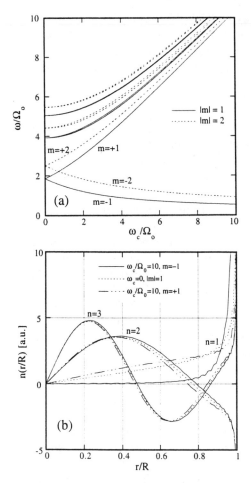

**Fig. 6.26.** (a) Magnetoplasmon-dispersion and (b) radial-induced density in circular electron disks, calculated in the framework of the hydrodynamical theory of [6.30]. With increasing $B$ the modes with negative magneto-dispersion $\omega_{1,-m}$ are localized at the edge (edge-magnetoplasmon, EMP). Fast oscillations are of numerical origin. The curves are normalized to $\int_0^1 dx\, x |n(x)| = 1$

Globally, $|\ell|$ decreases with an increasing magnetic field, so that, eventually, the regime $|\ell| \ll h$ is reached. The EMP-frequency is then given by

$$\omega_{1,-m} = -\frac{m\sigma_{xy}}{\varepsilon\varepsilon_0 d}\left(\ln\frac{d}{mh} + C\right), \quad |\ell| \ll h. \tag{6.34}$$

The "form factor" $C$, a dimensionless constant of the order of 1, depends on the specific form of the density profile,

$$C = -\gamma + \int\limits_{-\infty}^{\infty} d\xi \frac{df(\xi)}{d\xi} \ln\frac{1}{|\xi|}, \tag{6.35}$$

where $\gamma = 0.577\ldots$ is Euler's constant. The form factor is calculated as $C = \ln 4 - \gamma = 0.809$ for model $A$ and $C = 2 - \gamma = 1.423$ for model $B$. Thus, for $|\ell| \ll h$, the EMP frequency is given in terms of edge profile parameters, in particular the lateral depletion length, which can consequently be extracted from the measured EMP dispersion. For the case $|\ell| \gg h$, i.e., for smaller magnetic fields, the step-like profile represents a good approximation, and the EMP frequency is determined by the intrinsic EMP localization (6.33),

$$\omega_{1,-m} = -\frac{m\sigma_{xy}}{\varepsilon\varepsilon_0 d} \left( \ln \frac{d}{m|\ell|} + 1 \right), \quad h \ll |\ell| \ll d. \tag{6.36}$$

Earlier microwave experiments on electron disks have been interpreted by means of this formula [6.50, 51]. For still smaller $B$ such that $|\ell|$ becomes comparable to the disk diameter $d$, the above analysis is no longer valid.

In Fig. 6.25b we compare the experimental resonance positions of the $\omega_{1,-1}$ branch with the above dispersion. In the regime $|\ell| \approx h$ we suitably interpolate between (6.34, 36). With the Drude expression (6.13) for the conductivity tensor using the sample parameters, and with $\varepsilon = (\varepsilon_{\mathrm{GaAs}} + 1)/2 = 6.8$, the lateral depletion length is found from a best fit to the data as $h = 0.35\,\mu$m (model $A$) or $h = 0.6\,\mu$m (model $B$). Comparable values have been obtained for other disk systems and for completely different experimental setups as discussed in Sect. 6.4.3. The accuracy of the determination of $h$ by this method is – for a given model – of the order of 20% [6.52]. The differences in the theoretical density profiles used arise from specific assumptions of the extension of the edge surface charge. In model $A$ the edge of the 2DES is supposed to arise from a negatively biased gate so that the external charge seen by the 2DES is distributed over the gate, whereas model $B$ assumes a line charge at the side of an etched mesa in the plane of the 2DES. With an improved knowledge of the edge electrostatics of the real system the accuracy of our method could therefore be substantially enhanced. The lateral depletion length has also been determined from magneto-resistance measurements in quantum wire arrays exploiting the theory of weak localization, yielding, as well, values of the order of 0.5 μm [6.53]. Moreover, it is seen from Fig. 6.25b that the dispersions using the rectangular and the ellipsoidal profile predict too high and too low magnetoplasma frequencies, respectively.

The validity of the Drude approximation entering the above analysis is rather unexpected for the high magnetic field regime addressed here, but is justified experimentally by the absence of quantum oscillations in the dispersion as well as in the linewidth. Presumably, the inhomogenous density distribution at the edge averages out any oscillatory character of the magneto-conductivity. This is distictly different from EMP studies in the RF-frequency range where dispersion and linewidth are found to oscillate with the magnetic field [6.54]. In this case, the length $\ell$ can assume values larger and smaller than $h$ depending on whether the filling factor is noninteger or integer (resp. fractional for low temperatures [6.55]). Here, $\ell$ is always smaller than the lateral depletion length (for the magnetic fields in question) so that a detailed knowledge of $\sigma_{xx}$ is, in fact, not needed for the analysis of $\omega(B)$. The absence of any influence of the quantization of $\sigma_{xy}$ on the EMP dispersion in our experimental systems can, however, not be explained within the above model.

b) Rings. The set of plasma oscillations for geometries that deviate from the disk shape can be expected to be distinctly different. This is first demonstrated for a 2DES confined in a ring geometry [6.56, 57]. Such a system may alternatively, be considered as a disk with a repulsive scatterer, also known as an antidot [6.58–60], at its center. The observed resonances indeed exhibit features of the spectra of both dot and antidot systems. Which description prevails, depends on the aspect ratio of the ring and on the magnetic field strength. Figure 6.27 shows the transmission amplitude for rings with outer and inner diameters of $d = 50\,\mu m$ and $d_i = 12\,\mu m$, respectively, in the following referred to as "broad rings". Around $B = 0$, the observed resonances, which are plotted against magnetic field in Fig. 6.27a, are reminiscent of the modes in circular disks (Fig. 6.25). The set of modes is here labeled as $\omega_{n\pm}$, $n = 0, 1, 2, \ldots$, where $n$ is the number of nodes in the radial direction. As before, the other azimuthal modes are not excited in the uniform external high-frequency field of the waveguide setup. With this notation the ring resonances $\omega_{n\pm}$ correspond to the disk resonances $\omega_{n+1,\pm1}$.

Owing to the broadband tunability of our analyzer we are able to measure the magnetoplasma resonances not only in magnetic field sweeps at a fixed frequency (Fig. 6.27a, b), but also in frequency sweeps at a fixed magnetic field (Fig. 6.27c), though in a limited frequency range and with reduced resolution in amplitude and phase. While the resonance positions can usually be equally well extracted from both type of measurements, the direct determination of the linewidth $\Delta\omega$ is possible only in frequency sweeps. Magnetic field sweeps yield the width only indirectly via the magneto-dispersion, $\Delta\omega \approx (\partial\omega/\partial B)\Delta B$. Further, it is evident that frequency sweeps are indispensable if the resonance position happens to be independent of the magnetic field, as it is indeed the case for certain branches at particular values of $B$ (branches $\omega_{0+}$ and $\omega_{1+}$ in Fig. 6.28). Because of insufficient signal strength such resonances could, however, not be resolved here. The capability of performing frequency sweeps is also of particular value if the oscillator strength is rapidly varying with the magnetic field, as will be seen in Sect. 6.4.2d.

The most prominent difference of the ring resonances, as compared to the resonances in a simple disk, is the behavior of the upper branch of the lowest doublet, the $\omega_{0+}$ mode. Near $B = 0$ this branch exhibits a positive magneto-dispersion, as expected for the upper branch of a circular disk, where it would approach cyclotron resonance with increasing $B$. Here, however, above $B \approx 0.25\,\text{T}$, the frequency of the $\omega_{0+}$ branch decreases with increasing magnetic field; at the same time its oscillator strength decreases strongly. Above $B \approx 1\,\text{T}$ the resonance positions are well described by the EMP dispersion (6.36) if $d$ is set to the inner diameter of the ring, as reflected in the solid line in Fig. 6.28a. The high-field tail of the $\omega_{0+}$ mode thus constitutes an antidot EMP, localized at the inner boundary of the ring. The $\omega_{0-}$ mode is

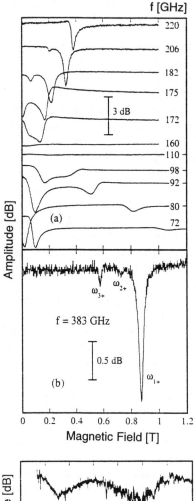

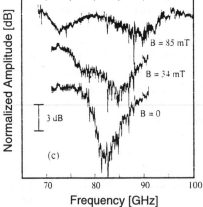

**Fig. 6.27 a–c.** Amplitude of the millimeter wave transmission for the array of electron rings with diameters $d = 50\,\mu$m and $d_i = 12\,\mu$m, referred to as wide rings in the text. (**a**) and (**b**) show data at fixed frequency $f$ vs magnetic field; (**c**) shows the normalized transmission vs frequency in the vicinity of zero magnetic field

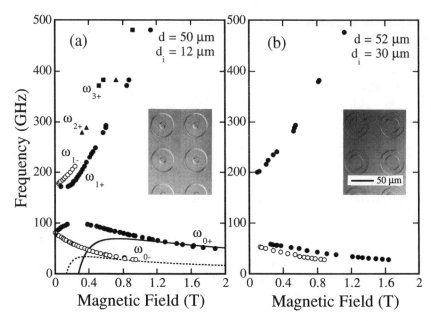

**Fig. 6.28.** Frequencies of the observed magnetoplasma resonances for two different ring geometries, (**a**) wide rings, (**b**) narrow rings (parameters are given in the text). In regions where the dispersion is flat the resonance positions cannot be resolved in our magnetic field sweep experiments ($\omega_{0+}$ and $\omega_{1+}$ branches). The curves in (**a**) are dispersions calculated from (6.36) for $d = 12\,\mu m$ (*solid line*) and $d = 50\,\mu m$ (*dashed line*). The insets show micrographs of the corresponding sample surface

the usual disk EMP at the outer edge. This interpretation is substantiated by the observations in the narrow rings with the same outer, but larger inner diameter (Fig. 6.28b). Here the frequency of the $\omega_{0+}$ mode has decreased, whereas the frequency of the outer EMP remains almost unchanged (except near $B = 0$, where the modes are not localized). The ratio of the measured $\omega_{0+}$ frequencies for the two rings is precisely reproduced by (6.36) for magnetic fields above $\approx 1\,T$. At high $B$, fits using (6.34) yield the same values for the lateral depletion length as the experiments on simple disks.

Contrary to experiments on disks, the splitting of the $\omega_1$ mode is also visible and is considerably larger than one would expect for the corresponding disk-resonances (Fig. 6.26). The lower branch of this mode dominates the spectrum at larger $B$ and is therefore a CR-like resonance, connected with right-handed electron trajectories, so that it is to be labeled as $\omega_{1+}$. This contrasts with the situation in simple disks where $\omega_{n+} > \omega_{n-}$. No doublets are observed for the $\omega_2$ and $\omega_3$ modes, presumably because of insufficient oscillator strength of their upper branches, $\omega_{2-}$ and $\omega_{3-}$. Classical calculations of the magnetoplasma spectrum in rings well agree with the experimental findings and confirm the above identification of the observed modes [6.31, 32, 61].

The vector measurement of the MMW transmission reveals an important detail of the dispersion which would not have been obtained with scalar detection. At the frequencies $f = 92, 98$ and $172\,\mathrm{GHz}$, the polar diagram, Fig. 6.29, shows two arc segments with clockwise and counterclockwise sense of rotation indicating a resonance with positive and one with negative magneto-dispersion, as discussed before. Such signals originate from the non-monotonic magneto-dispersion of the corresponding branch. In particular, the 172 GHz curve shows that the $\omega_{1+}$ branch has a negative magneto-dispersion near $B = 0$, although data are only available for one fixed frequency for the spectral range of interest here (at $175\,\mathrm{GHz}$ one observes the doublet $\omega_{1\pm}$, since both resonances exhibit positive magneto-dispersion). The vector fit procedure introduced in Sect. 6.3.3 allows the resonances to be analyzed quantitatively even if only a small fraction of the resonance curve is swept through in the experiment.

We also note that the excitations in a narrow ring (Fig. 6.28b) can be understood as 1D magnetoplasmons with wave vector $q = 2/\bar{d}$, where $\bar{d}$ is the mean diameter of the ring. From the calculations of *Eliasson* et al. [6.33] we obtain $\omega_0 = 57\,\mathrm{GHz}$ and $\omega_1 = 200\,\mathrm{GHz}$ for our sample parameters for

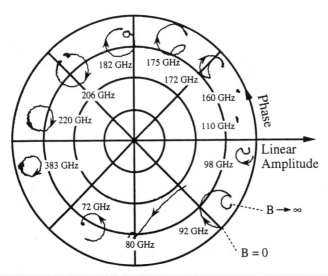

**Fig. 6.29.** Polar diagram (amplitude vs phase) corresponding to the amplitude data displayed in Fig. 6.27a,b. Curve parameter is the magnetic field. Full coverage of a resonance in a sweep results in a closed circle. A partial sweep through a resonance leads to an arc segment. This is the case if a resonance is close to $B = 0$ or if a branch is not a monotonic function of $B$ and is swept through twice. The sense of rotation along the curves with an increasing magnetic field is clockwise or counterclockwise for modes with positive resp. negative magneto-dispersion. In particular, the signals at 92, 98 and 172 GHz each exhibit one mode with positive and one with negative dispersion, whereas, e.g., for $f = 175\,\mathrm{GHz}$ both of the two resonances belong to modes with positive dispersion

zero magnetic field, which is in good agreement with the experimental values extrapolated to $B = 0$. Even the spectrum in wide rings is reminiscent of the wire structure: The oscillator strength of the highest observed ($\omega_3$) mode is larger than the strength of the $\omega_2$ mode, which at first appears surprising as it is different for simple disks (Fig. 6.24). In a straight strip, however, only the modes with an odd number of nodes possess a finite dipole moment. Altering the symmetry of the strip by "bending" it to a ring leads to a nonzero moment for the even modes as well, yielding a small but discernable signal for the $\omega_2$ mode in the wide-ring geometry (Fig. 6.27b). For the narrow rings the excitation of higher order modes is obviously too weak to be observable for both even and odd modes.

c) **Elliptic Disks.** In the following sections we discuss plasma excitations in systems with only two-fold symmetry, and begin with the particular case of elliptic disks in order to facilitate the comparison with theory. The anisotropic response of such structures manifests itself in a lifting of the degeneracy of the plasma resonances at $B = 0$, a dependence of the oscillator strength on the polarization of the incident MMW and a nonzero dipole moment for certain modes $|m| > 1$ [6.62].

The two fundamental modes can again be described within the ellipsoidal model, approximating the 2D ellipse by a thin 3D ellipsoid with semi-major and semi-minor axes, $a$ and $b$, along the $x$- and $y$-direction, respectively. We consider only the case of weak damping, $\omega\tau \gg 1$. The depolarization factors $L_x$ and $L_y$ entering the equation of motion (6.23) are no longer equal, giving rise to two $B = 0$ frequencies

$$
\begin{aligned}
\omega_x^2 &= \omega_{\text{p,3D}}^2 L_x \approx \frac{3n_s e^2 b}{4\varepsilon\varepsilon_0 m^* a^2} \frac{K(\eta) - E(\eta)}{\eta^2}, \\
\omega_y^2 &= \omega_{\text{p,3D}}^2 L_y \approx \frac{3n_s e^2}{4\varepsilon\varepsilon_0 m^* b} \frac{E(\eta) - (1 - \eta^2)K(\eta)}{\eta^2}.
\end{aligned}
\tag{6.37}
$$

Here, $K$ and $E$ are the complete elliptic integrals of the first and second kind and $\eta^2 = 1 - b^2/a^2$ is a measure of the eccentricity of the ellipse. For $b \to a$ one recovers the resonance frequency of the thin oblate spheroid, $\omega_x = \omega_y = \omega_0$, (6.27). The magnetic field dependence of the plasma frequencies (6.37) is given by

$$
\omega_\pm^2 = \tfrac{1}{2}\left(\omega_x^2 + \omega_y^2 + \omega_c^2\right) \pm \sqrt{\tfrac{1}{4}\left(\omega_x^2 + \omega_y^2 + \omega_c^2\right)^2 - \omega_x^2 \omega_y^2},
\tag{6.38}
$$

and for the conductivity we obtain

$$
\sigma^{\text{ext}} = \frac{i\omega\sigma_0/\tau}{\left(\omega^2 - \omega_x^2 + i\omega/\tau\right)\left(\omega^2 - \omega_y^2 + i\omega/\tau\right) - \omega^2\omega_c^2}
$$
$$
\Big/ \left( \begin{array}{cc} \omega^2 - \omega_y^2 + i\omega/\tau & -i\omega\omega_c \\ i\omega\omega_c & \omega^2 - \omega_x^2 + i\omega/\tau \end{array} \right).
\tag{6.39}
$$

Figure 6.30 compares experimentally observed frequencies in their dependence on the magnetic field for both polarizations. The dimensions of

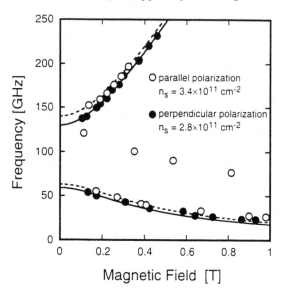

**Fig. 6.30.** Magnetic field dispersion of plasma resonances in $116 \times 40\,\mu\text{m}$ elliptic electron disks. The lines are calculated from (6.38) fitted to the experimental resonance positions for the parallel (---) and perpendicular (—) geometry, respectively, with parameters as given in the text

the elliptic disks studied here are $2a = 116\,\mu\text{m}$ and $2b = 40\,\mu\text{m}$. For the case of parallel polarization the best fit to the resonance positions according to (6.38) is obtained with $\omega_x/2\pi = 63\,\text{GHz}$ and $\omega_y/2\pi = 140\,\text{GHz}$ using $m^* = 0.069\,m_\text{e}$. These frequencies, which are thus extrapolated from the experimental resonance positions at finite $B$, are well consistent with (6.37) which yields $\omega_x/2\pi = 64\,\text{GHz}$ and $\omega_y/2\pi = 143\,\text{GHz}$ for $n_\text{s} = 3.4 \times 10^{11}\,\text{cm}^{-2}$ and $\varepsilon = 6.8$. The density is obtained from Shubnikov-de Haas oscillations visible in the microwave signal for lower frequencies at $T = 2\,\text{K}$. Thermal cycling of the sample between the two different polarization measurements is the origin of the somewhat lower density, $n_\text{s} = 2.8 \times 10^{11}\,\text{cm}^{-2}$, that prevails for the measurements in perpendicular polarization. This results in correspondingly lower plasma frequencies, which, again, are in good agreement with (6.37). Both modes appear in both polarizations, because they are measured at a finite magnetic field; they merely exhibit a polarization-dependent oscillator strength. Complete anisotropy would only be obtained for $B = 0$.

The additional branch with negative magneto-dispersion, observed only in parallel polarization (Fig. 6.30), constitutes the mode $\omega_{1,-3}$, i.e., a higher azimuthal mode [6.63]. Here, the radial and azimuthal mode indices, $n$ and $m$, are generalized to the elliptic geometry. Contrary to the corresponding mode in a circular disk, this resonance possesses a nonzero dipole moment giving rise to a finite oscillator strength. We recall, however, that for parabolic confinement – even if anisotropic – only the fundamental modes are excited in the external dipole field. The higher modes with finite dipole moment appear because the lateral confinement is strongly nonparabolic here, so that Kohn's theorem does not apply. In the limit of a high-eccentricity ellipse, the two branches with negative magneto-dispersion $\omega_{1,-1}$ and $\omega_{1,-3}$ can approximately be interpreted as 1D plasmons with wavevectors $q \approx \pi/2a$ and $q \approx 3\pi/2a$, respectively. In this sense the resonances in an elliptic disk interpolate the electronic excitations in zero- and one-dimensional systems.

**d) Coulomb Coupling in Disk-Arrays.** In a densely packed array the plasma frequencies may differ considerably from the response of a single disk. When a plasma oscillation is excited in such a system, the depolarization field within a given disk is superposed with the Lorentz-field of the neighboring disks. The interdisk interaction can unambiguously be demonstrated and quantified for a system of circular disks in a rectangular lattice with strongly differing periods [6.64]. This constitutes a system with only twofold symmetry and therefore anisotropic conductivity tensor, $\sigma_{yy} \neq \sigma_{xx}$. Since the conductivity of the individual circular disk is isotropic, an experimentally observed anisotropy is entirely due to the local field correction of the lattice. The local field depends on the orientation of the plasma oscillation with respect to the axes of the lattice so that two fundamental plasma frequencies occur at $B = 0$. The size of the gap is a measure of the interdisk coupling strength.

The anisotropy of the system is best demonstrated by frequency sweeps at zero magnetic field as displayed in Fig. 6.31. As for the rings, the observation of the resonances in the frequency sweep is also possible here because of sufficient bandwidth ($\approx 25$ GHz) in the spectral range of interest. Depending on the polarization, only the upper or the lower resonance is observed at $B = 0$. Passing from parallel to perpendicular polarization (i.e., MMW

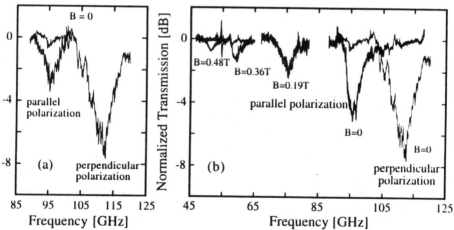

**Fig. 6.31 a,b.** Transmission of the disk array of Fig. 6.32 vs frequency for parallel and perpendicular polarization (i.e., millimeter wave polarized along or perpendicular to the short period of the disk array, respectively), revealing the strong anisotropy of the electron system. (**a**) Sample mounted in perpendicular polarization; the signal in parallel polarization is obtained by turning both the sender and detector diode. (**b**) The different polarizations are obtained by turning the sample. Note the different oscillator strength for parallel polarization as compared to (**a**). Further sweeps at increasing magnetic fields demonstrate the behavior of the lower branch

polarized parallel or perpendicular to the short lattice axis) can be achieved by turning sender and detector using appropriate tapered transitions, or by turning and remounting the sample itself. In the first case, the waveguide is operated in the modes $\mathrm{TE}_{no}$ and $\mathrm{TE}_{om}$, respectively. At 100 GHz the largest possible indices for our Q-band waveguide are $n = 3$ and $m = 1$. Only odd TE modes are expected to be excited for the usual sender diode geometries. Mixing between polarizations could be kept below $-20\,\mathrm{dB}$. The disadvantage of turning the MMW polarization lies in different leakage transmissions, if, as in the present case, the sample does not cover the whole waveguide section. The relative resonance signals are, therefore, not comparable (Fig. 6.31a). On the other hand, turning the (square shaped) sample ensures comparable oscillator strengths, but may, due to the necessary thermal cycle, lead to different carrier densities as discussed in the preceding section. We have not attempted to construct a setup with in-situ rotation of the sample. A possibly lower density may, however, be raised to the desired value by means of the persistant photo effect. Figure 6.31b thus shows curves which correctly represent the behavior of the system in both respects, resonance positions and oscillator strengths. Figure 6.32 exhibits the magnetic field dependence of the observed resonance positions, as well as a micrograph of the sample surface. As before, most of the data have been obtained from magnetic field sweeps at a fixed frequency (not shown here).

To model the interdisk coupling we calculate the Lorentz-correction in summing up the dipole fields of all disks, with respect to a given one. For a given wave vector $\boldsymbol{k}$ this interdisk field is given by:

$$\boldsymbol{E}_{\mathrm{inter}}\left(\boldsymbol{k}\right) = \frac{1}{\varepsilon\varepsilon_0 a^3}\xi\left(\boldsymbol{k}\right)\boldsymbol{p}\left(\boldsymbol{k}\right), \tag{6.40}$$

where $\boldsymbol{p}$ is the induced dipole moment of a disk and $\xi$ the diagonal "lattice tensor" with the diagonal elements

$$
\begin{aligned}
\xi_x\left(\boldsymbol{k}\right) &= \frac{1}{4\pi}\sum_{mn}{}'\frac{2n^2 - \beta^2 m^2}{\left(n^2 + \beta^2 m^2\right)^{5/2}}\mathrm{e}^{i\boldsymbol{k}\boldsymbol{r}_{nm}}, \\
\xi_y\left(\boldsymbol{k}\right) &= \frac{1}{4\pi}\sum_{mn}{}'\frac{2\beta^2 m^2 - n^2}{\left(n^2 + \beta^2 m^2\right)^{5/2}}\mathrm{e}^{i\boldsymbol{k}\boldsymbol{r}_{nm}}.
\end{aligned}
\tag{6.41}
$$

Here, $\beta = b/a$ and $\boldsymbol{r}_{nm} = (na, mb)$, where $a$ and $b$ are the short and long period lattice constants along directions $x$ and $y$, respectively. The sum extends over all disks excluding $n = m = 0$. Experimentally we have $k \approx 0$ and $\beta = 2$, which yields $\xi_x = 0.3825$ and $\xi_y = -0.0604$ if the lattice is assumed to be infinite. Lattice sums for other geometries and values of $k$ have been calculated elsewhere. Within the ellipsoidal model, where the polarization is homogeneous, $\boldsymbol{p} = P V$, the Lorentz field (6.40) and the intradisk depolarization field $-\mathbf{L}\boldsymbol{P}/\varepsilon\varepsilon_0$, can be combined to an "effective depolarization field" with an effective depolarization tensor $\mathbf{L}_{\mathrm{eff}} = \mathbf{L} - V\boldsymbol{\xi}/a^3$. The response thus becomes formally identical to the response in elliptic electron disks, (6.38, 39), with the $B = 0$ plasma frequencies

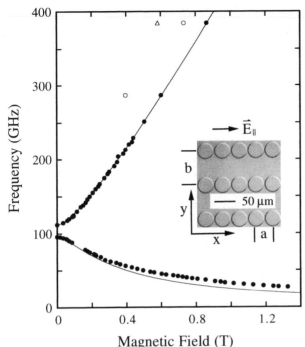

**Fig. 6.32.** Resonance positions of the rectangular disk array (parameters are given in the text). The gap at zero magnetic field is a measure of the Coulomb interaction between disks. The *solid lines* reflect the magnetic field dispersion of the dipole model, (6.38, 42), with the $B = 0$ resonance frequencies and $m^* = 0.068\,m_e$ as fit parameters. *Open symbols* denote higher modes. The *inset* shows a micrograph of the sample surface

$$\omega_{x,y}^2 = \omega_0^2 \left(1 - \frac{2d^3}{3a^3}\xi_{x,y}\right). \tag{6.42}$$

Equation (6.42) is the starting point for some physically interesting quantities. The absolute shift $\omega_0^2 - \omega_{x,y}^2$ constitutes the measure of the absolute Coulomb interaction between disks, where the noninteracting plasma frequency $\omega_0$ can be interpolated from the measured $\omega_x$ and $\omega_y$ as the weighted mean, $\omega_0^2 = (\xi_x\omega_y^2 - \xi_y\omega_x^2)/(\xi_x - \xi_y)$. The relative splitting,

$$\frac{\omega_y^2 - \omega_x^2}{\omega_0^2} = \frac{(\xi_x - \xi_y)\,V}{L_{||}a^3} \tag{6.43}$$

measures the ratio of inter- to intradisk field. It is given in terms of geometrical quantities only, independent of the intrinsic sample parameters $n_s$, $m^*$ and $\varepsilon$. The absolute splitting

$$\omega_y^2 - \omega_x^2 = (\xi_x - \xi_y)\,Ne^2/\varepsilon\varepsilon_0 m^* a^3 \tag{6.44}$$

is independent of the shape of the individual disks, as long as their fourfold symmetry is preserved. This quantity allows for a determination of the mean number $N$ of electrons in a disk.

The measured $B = 0$ frequencies and the derived quantities (6.42–44) are in reasonable agreement with the above dipole approximation. Deviations have been attributed to the effect of higher multipole moments. The open symbols in Fig. 6.32 correspond to plasmons with higher radial mode index. Higher order modes are less affected by interdisk interaction and can here be described in the single-disk approximation. However, for further reduced spacing between disks, higher azimuthal modes become dipole-active due to the mutual interaction, in further analogy with the observation in elliptic disks. The deviation of the magnetic field dispersion of the lower branch from the ellipsoidal model has already been discussed in Sect. 6.4.2a.

With the above results we can also correct the plasma frequency in the square lattice of disks of Sect. 6.4.2a. For $\beta = 1$ and $k = 0$ (6.41) simplifies to

$$
\xi_x = \xi_y = \frac{1}{2\pi} \left( \sum_{m,n=1}^{\infty} \frac{1}{(m^2 + n^2)^{3/2}} + \sum_{n=1}^{\infty} \frac{1}{n^3} \right)
$$

$$
= 0.360.
$$

(6.45)

The response of the square lattice is isotropic so that the magnetoplasma spectrum is qualitatively similar to the noninteracting case. Quantitatively, however, the inclusion of the Lorentz correction according to (6.45) can considerably improve the agreement with the experiment.

Within the present model, maximum coupling for circular disks at $k = 0$ would occur for the hexagonal lattice with close packing, yielding a relative shift of $\left( \omega_0^2 - \omega_x^2 \right) / \omega_0^2 \approx 0.29$. Larger coupling has been observed in systems of elongated disks [6.63]. In suitable quantum-dot structures containing a well-defined, small number of electrons, the interdisk coupling is expected to overcome the intradisk restoring forces, so that the dots polarize spontaneously [6.65]. While in a classical "metallic" system such a scenario is principally not possible [6.66], the ferro- or antiferroelectric phase transition of quantum-dot arrays [6.67] resembles the spontaneous ordering of the 2DEG into a Wigner crystal [6.68], which has been observed experimentally for electrons on the surface of liquid helium [6.69]. It has even been suggested that the mutual polarization in specially designed quantum-dot structures may be exploited to realize new quantum devices based on the principle of cellular automata [6.70].

### 6.4.3 Towards Millimeter-Wave Spectroscopy of Mesoscopic Systems

**a) Coupling via Coaxial Cable and Microstripline.** The spectroscopy of single, small electron systems is of great interest since it reduces fabrication requirements and eliminates inhomogeneous broadening. Until very recently, experiments on individual quantum dots have only been reported for optical wavelengths. However, recent photoconductivity studies of quantum point contacts in the far-infrared [6.71] and on individual quantum dots at centimeter wave frequencies [6.72] show that the necessary focusing techniques to couple long wavelength radiation into sub-μm structures are available, so that transmission (or reflection) experiments on single dots should also, in principle, be possible. Comparatively large disks have been investigated in the RF and microwave regime employing microstripline techniques [6.73, 52]. Figure 6.33 shows microwave transmission data for circular disks with diameters $d = 100\,\mu$m and $40\,\mu$m. The microwave is guided by silver microstriplines, evaporated onto the sample on opposite sides of each disk and designed to have an impendance of $50\,\Omega$. The spacing between the ends of the striplines and the mesa edges was chosen as $d/2$, which is sufficiently close to keep the RF power seen by the disk comparable with the direct crosstalk between striplines but still far enough to leave the plasma oscillations undisturbed to a good approximation. Coaxial cables connect the sample to the outside. The

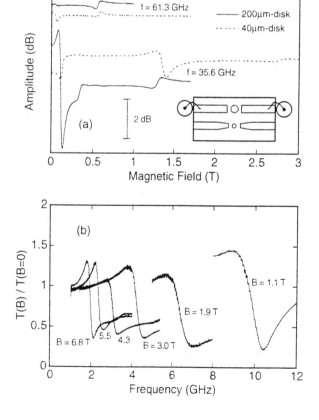

**Fig. 6.33.** (a) Transmitted microwave amplitude at fixed frequency $f$ vs magnetic field for single electron disks. The *solid* (*dotted*) lines display the signal for the case that the coaxial leads are connected to the striplines leading to the 200 μm (40 μm) disk; the other disk is excited by crosstalk. The *inset* schematically shows the sample setup (b) Normalized linear transmission amplitude vs frequency at fixed magnetic fields for the 200 μm disk

microwave transmission is recorded either with the help of a RF-synthesizer employing homodyne detection ($f < 8\,\mathrm{GHz}$), or by means of the analyzer introduced in Sect. 6.1 ($f > 8\,\mathrm{GHz}$). In magnetic field sweeps at fixed frequency the setup allows a range from below $1\,\mathrm{GHz}$ up to more than $60\,\mathrm{GHz}$ to be covered (Fig. 6.33a). In variable frequency measurements (Fig. 6.33b), the disk resonances have been observed up to about $12\,\mathrm{GHz}$.

From their negative magneto-dispersion (Fig. 6.34) the observed modes are identified as EMPs $\omega_{1,-1}$ and $\omega_{1,-2}$. The $200\,\mu\mathrm{m}$ disk also exhibits the beginning of the CR-like mode $\omega_{1,+1}$. Higher azimuthal modes ($m > 1$) are excited here, because the external microwave field is spatially inhomogeneous. The fits to the EMP theories introduced in Sect. 6.4.2a confirm the influence of the lateral depletion length $h$ on the EMP dispersion and yield values for $h$ that are comparable to the ones obtained from the MMW measurements on the $11.5\,\mu\mathrm{m}$-disks. Contrary to the disk arrays, the extrapolated $B = 0$ plasma frequency of both disks is in good agreement with (6.32). Further, the comparatively weak coupling of the microwave into the electron disk allows the (unloaded) EMP linewidths to be measured directly. They are found to decrease roughly proportionally to the EMP frequencies, $\Delta\omega/\omega \approx \Delta B/B \approx 0.3$, in reasonable agreement with the theory of Volkov and Mikhailov using a step-like density. The influence of a smooth edge profile on the EMP linewidths is, at present, not known quantitatively. Note that the ellipsoidal model predicts a $B$-dependent width $\Delta\omega/\omega \sim 1/B$, (6.30).

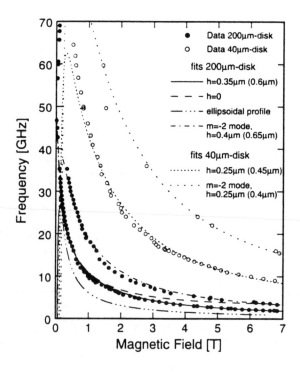

**Fig. 6.34.** Positions of the observed magneto-plasma resonances for two disk diameters together with magnetic field dispersions calculated from (6.34, 36) using the lateral depletion length $h$ as fit parameter, as discussed in Sect. 6.4.2.1. Again, the two different values for $h$ are obtained for model $A$ (model $B$), respectively, which describe the electrostatics at the edge of the 2DEG. Also shown are dispersions for both an abrupt and an ellipsoidal density profile

**b) Focussing with Diaphragms.** The limited frequency range of the above technique prohibits its direct application to the spectroscopy of quantum dots. Even if the coaxial leads are replaced by waveguides, serious problems persist concerning the waveguide to stripline transition for the frequency range in question. Also, in order to suppress radiation from the simple microstripline at millimeter wavelength, one would have to use a triplate stripline. Even then, direct crosstalk between stripline ends would eventually bury the dot signal in the "leakage wave" bypassing the quantum dot.

Leakage is easily eliminated by suitable diaphragms, which can be fabricated directly onto the sample surface by conventional lithographic techniques. The residual transmission must merely stay well above the noise floor of the detection system. However, since the radiation couples to the plasma-excitation via the electric field, only capacitive diaphragms should be considered. Unfortunately, the most natural solution, a small circular aperture in the waveguide cross section, represents an inductive impedance. The electric field is practically short-circuited and propagation through the hole is accomplished via the magnetic part of the MMW field [6.74]. In the following, we show that considerable focusing can also be obtained by a narrow slit, which constitutes a capacitive obstacle.

The final slit dimensions are obtained by combining a tapered waveguide with a silver diaphragm on the sample. The tapered transition reduces the waveguide dimensions from $5.69\,\text{mm} \times 2.84\,\text{mm}$ to $2.7\,\text{mm} \times 0.1\,\text{mm}$. The sample is placed on the backside of the tapered piece. As compared to a simple slit with the same dimensions, the tapered transition enhances the transmission substantially. Without further reduction in size, such a tapered slit allows plasma resonances in disks with diameters down to $20\,\mu\text{m}$ to be observed in the transmission. For larger disk diameters, the setup can be used to excite plasma modes with zero dipole moment , since the MMW field near the edge of the slit is strongly inhomogenous – similar to the stripline mounting discussed in the preceding section. A method to excite quadrupole modes in quantum dots is theoretically discussed in [6.75].

Figure 6.35 shows data for a $200\,\mu\text{m}$-disk which is placed asymmetrically in the slit opening. First, the fundamental plasma resonance is found to have substantially larger amplitude than for the conventional transmission setup. Secondly, higher order modes are excited, which, from the plot of the resonance position against magentic field, are identified as azimuthal modes, $\omega_{1,\pm m}$. Here we observe modes up to the fourth harmonic. Due to screening of the nearby metal, the absolute frequencies are somewhat lower than calculated in Sect. 6.4.2a (6.32) and Fig. 6.26. Comparison of the measured resonance positions with the theoretical magneto-dispersion in Fig. 6.36 shows that the effect can approximately be described by an increased effective dielectric constant, though in principle a spatially inhomogenous metallization would require a more elaborated theory.

Additional diaphragms on the sample surface further enhance leakage rejection and allow for an investigation of still smaller systems. The spectra of

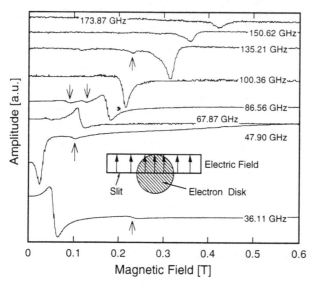

**Fig. 6.35.** Amplitude of transmitted millimeter wave of a 200 μm disk, asymmetrically positioned at the end of a narrow waveguide as sketched in the inset. As in the microstripline setup of Fig. 6.39, the inhomogeneous field distribution at the location of the disk leads to the excitation of modes with higher azimuthal index (indicated)

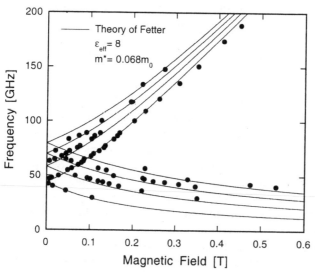

**Fig. 6.36.** Resonance positions from the measurements of Fig. 6.35 together with fits calculated from the theory of [6.30] (see Fig. 6.26)

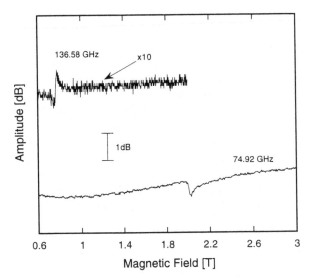

**Fig. 6.37.** Transmitted amplitudes for a single 9 μm disk vs magnetic field at different frequencies. The millimeter wave is focused onto the disk by means of a tapered waveguide in conjunction with a diaphragm having an opening of 30 μm which is evaporated onto the sample

Fig. 6.37 exhibit the EMP resonance $\omega_{1,-1}$ of a single dot having a diameter of 9 μm. On both sides of the dot a 0.5 μm-thick silver layer was evaporated, leaving a slit width of 30 μm. The thickness of the metal must be well above the skin depth in the spectral range of interest. The resonances could be observed at frequencies ranging from 60 to 150 GHz. These results are very encouraging as to the problem of the spectrocopy of a true few-electron system, since the presented setup is not yet optimized. The application of differential measuring techniques can be expected to enhance the resolution of the apparatus by two orders of magnitude. This translates roughly into a possible reduction of the dot diameter by a factor of 10. The methods may involve a microwave bridge or the modulation of dot parameters (magnetic field modulation, electron density modulation via Schottky gates). Also, because of the increased resonance frequencies in smaller dots – usually in the submillimeter regime – the focusing problem is slightly relieved. For our diaphragm technique this means that the waveguide dimensions and, hence, the slit length, may be reduced to $\lambda_0/2$. Finally, cavity or Fabry-Perot setups may be employed in quantum-dot spectroscopy; however, with the disadvantage of losing the possibility to continuously tune the frequency.

## 6.5 Conclusion

The examples illustrated above conclusively show that linear millimeter wave spectroscopy is aided tremendously by the introduction of a millimeter wave network analyzer that is capable of performing sensitive experiments at frequencies well above 100 GHz. In these experiments very low power levels are employed and sensitive detection is achieved via heterodyning. This is particularly important if one wants to study systems that are easily disturbed by stronger microwave fields. The mesoscopic semiconductor systems illustrated in the last chapter are such sytems as are quantum dots and quantum point contacts that are operated at low temperatures [6.72]. One can expect that for the study of high-frequency properties in such delicate quantum systems, the millimeter wave network analyzer discussed here will be extremely valuable.

## References

6.1     S.M. Sze (ed).: *High-Speed Semiconductor Devices* (Wiley, New York 1990)

6.2     J.-F. Luy, P. Russer (eds.): *Silicon-Based Millimeter-Wave Devices*, Springer Ser. Electron. Photon., Vol.32 (Springer, Berlin, Heidelberg 1994)

6.3     R.J. Bell: *Introductory Fourier Transform Spectroscopy* (Academic, New York 1972)

6.4     E. Batke, D. Heitmann: Infrared Phys. **24**, 189 (1984)

6.5     W. Hansen: In *Quantum Coherence in Mesoscopic Systems*, ed. by B. Kramer, J.E. Mooj, A. MacKinnon, R.A. Webb (Plenum, New York 1990) pp.23-41

6.6     J.P. Kotthaus: In *Frontiers of Optical Phenomena in Semiconductor Structures of Reduced Dimensions*, ed. by D.J. Lockwood (Kluwer, Dordrecht 1993) pp.245-264

6.7     P. Goy, M. Gross, J.M. Raimond, J.C. Buisson: Ann. Télécommun. **43**, 331 (1988)

6.8     R.K. Hoffmann: *Integrierte Mikrowellenschaltungen* (Springer, Berlin, Heidelberg 1983)

6.9     A. Wittlin, M.E. Boonman, J.C. Maan: Unpublished (1993)
        M.E.J. Boonman: Millimeter wave spectroscopy in high magnetic fields, Dissertation, University of Nijmegen (1998)

6.10    T. Ando, A.B. Fowler, F. Stern: Rev. Mod. Phys. **54**, 437 (1982)

6.11    D.C. Tsui, S.J. Allen Jr, R.A. Logan, A. Kamgar, S.N. Coppersmith: Surf. Sci. **73**, 419 (1978)

6.12    G. Abstreiter, J.P. Kotthaus, J.F. Koch, G. Dorda: Phys. Rev. B **14**, 2480 (1976)

6.13    J.F. Koch: Surf. Sci. **58**, 104 (1976)

6.14    K.W. Chiu, T.K. Lee, J.J. Quinn: Surf. Sci. **58**, 182 (1976)

6.15    R. Meisls, F. Kuchar: Z. Physik **67**, 199 (1987)

6.16    Xu Shanjia, Wu Xinzhang, P. Boege, H. Schäfer, C.R. Becker, R. Geick: Int'l J. Infrared and Millimeter Waves **14**, 2155 (1993)

6.17    F. Brinkop: Mikrowelleneigenschaften niederdimensionaler Elektronensysteme in AlGaAs/GaAs-Heterostrukturen, Dissertation, Universität München (1993)

6.18    F. Kuchar, R. Meisls, G. Weimann, W. Schlapp: Phys. Rev. B 33, 2965 (1986)
        F. Kuchar: Int'l J. Mod. Phys. B 3, 1129 (1989)
6.19    L.A. Galchenkov, I.M. Grodnenskii, M.V. Kostovetskii, O.R. Matov: Pis'ma Zh.
        Eksp. Teor. Fiz. 46, 430 (1987) [JETP Lett. 46, 542 (1987)]
6.20    O. Vieweger, K.B. Efetov: In High Magnetic Fields in Semiconductor Physics III, ed.
        by G. Landwehr, Springer Ser. Solid-State Sci. Vol. 101 (Springer, Berlin, Heidel-
        berg 1992)
6.21    F. Stern: Phys. Rev. Lett. 18, 546 (1967)
6.22    C.C. Grimes, G. Adams: Phys. Rev. Lett. 36, 145 (1976)
6.23    S.J. Allen Jr., D.C. Tsui, R.A. Logan: Phys. Rev. Lett. 38, 980 (1977)
        T.N. Theis, J.P. Kotthaus, P.J. Stiles: Solid State Commun. 24, 273 (1977)
6.24    A.L. Fetter: Ann. Phys. (N.Y.) 81, 367 (1973)
6.25    J.D. Jackson: Classical Electrodynamics (de Gruyter, Berlin 1983) p. 581
6.26    A.V. Chaplik: Surf. Sci. Rpts. 5, 289 (1985)
6.27    J. Dempsey, B.I. Halperin: Phys. Rev B 45, 1719 (1992)
6.28    V.A. Volkov, S.A. Mikhailov: Zh. Eksp. Teor. Fiz. 94, 217 (1988) [Sov. Phys.
        JETP 67, 1639 (1988)]
6.29    V.A. Volkov, S.A. Mikhailov: In Landau Level Spectroscopy, ed. by G. Landwehr,
        E.I. Rashba (Elsevier, Amsterdam 1991)
6.30    A.L. Fetter: Phys. Rev. B 33, 5221 (1986); and ibid. B 32, 7676 (1985)
6.31    F.A. Reboredo, C.R. Proetto: Phys. Rev. B 53, 12617 (1996)
        C.R. Proetto: Phys. Rev. B 46, 16174 (1992)
6.32    H.L. Cui, V. Fessatidis, O. Kühn: Superlattices and Microstructures 17, 173 (1995)
6.33    G. Eliasson, J.-W. Wu, P. Hawrylak, J.J. Quinn: Solid State Commun. 60, 41
        (1986)
6.34    E. Zaremba, H.C. Tso: Phys. Rev. B 49, 8147 (1994)
6.35    W. Kohn: Phys. Rev. 123, 1242 (1961)
6.36    L. Brey, N.F. Johnson, B.I. Halperin: Phys. Rev. B 40, 10647 (1989)
6.37    P. A. Maksym, T. Chakraborty: Phys. Rev. Lett. 65, 108 (1990)
6.38    W. Hansen, T.P. Smith III, K.Y. Lee, J.A. Brum, C.M. Knoedler, J.M. Hong,
        D.P. Kern: Phys. Rev. Lett. 62, 2168 (1989)
6.39    P.L. McEuen, E.B. Foxman, U. Meirav, M.A. Kastner, Y. Meir, N.S. Wingreen,
        S.J. Wind: Phys. Rev. Lett. 66, 1926 (1991)
6.40    R.L. Leavitt, J. W. Little: Phys. Rev. B 34, 2450 (1986)
6.41    V.I. Talyanskii: Zh. Eksp. Teor. Fiz 92, 1845 (1987) [Sov. Phys. - JETP 65, 1036
        (1987)]
6.42    B.A. Wilson, S.J. Allen Jr., D.C. Tsui: Phys. Rev. B 24, 5887 (1981)
6.43    S.A. Govorkov, M.I. Reznikov, B.K. Medvedev, V.G. Mokerov, A.P. Senichkin,
        V.I. Talyanskii: Pis'ma Zh. Eksp. Teor. Fiz. 45, 252 (1987) [JETP Lett. 45, 316
        (1987)]
6.44    V. Gudmundsson, R.R. Gerhardts: Phys. Rev. B 43, 12098 (1991)
6.45    S.J. Allen Jr., H.L. Störmer, J.C.M. Hwang: Phys. Rev. B 28, 4875 (1983)
6.46    D.B. Mast, A.J. Dahm, A.L. Fetter: Phys. Rev. Lett. 54, 1706 (1985)
        D.C. Glattli, E.Y. Andrei, G. Deville, J. Poitrenaud, F.I.B. Williams: Phys.
        Rev. Lett. 54, 1710 (1985)
6.47    D. Heitmann, J.P. Kotthaus: Phys. Today 46, 56 (June 1993)

6.48    D.B. Shklovskii, B.I. Shklovskii, L.I. Glazman: Phys. Rev. B **46**, 4026 (1992)

6.49    B.Y. Gelfand, B.I. Halperin: Phys. Rev. B **49**, 1862 (1994)

6.50    L.A. Galchenkov, I.M. Grodnenskii, M.V. Kostovetskii, O.R. Matov: Pis'ma Zh. Eksp. Teor. Fiz. **46**, 430 (1987) [JETP Lett. **46**, 542 (1987)]

6.51    F. Brinkop, C. Dahl, J.P. Kotthaus, G. Weimann, W. Schlapp: In *High Magnetic Fields in Semiconductor Physics III*, ed. by G. Landwehr, Springer Ser. Solid-State Sci., Vol. 101 (Springer, Berlin, Heidelberg 1992)

6.52    C. Dahl, S. Manus, J.P. Kotthaus, H. Nickel, W. Schlapp: Appl. Phys. Lett. **66**, 2271 (1995)

6.53    K.K. Choi, D.C. Tsui, K. Alavi: Appl. Phys. Lett. **50**, 110 (1987)

6.54    I. Grodnenskii, D. Heitmann, K. v. Klitzing: Phys. Rev. Lett. **67**, 1019 (1991)

6.55    M. Wassermeier, J. Oshinowo, J.P. Kotthaus, A.H. MacDonald, C.T. Foxon, J.J. Harris: Phys. Rev. B **41**, 10287 (1990)

6.56    C. Dahl, J.P. Kotthaus, H. Nickel, W. Schlapp: Phys. Rev. B **48**, 15480 (1993)

6.57    P.J.M. Peters, P.K.H Sommerfeld, S. van der Berg, P.P. Steijaert, R.W. van der Heijden, A.T.A.M. de Waele: Physica B **204**, 105 (1995)

6.58    D. Weiss, K. Richter, E. Vasiliadou, G. Lütjering: Surf. Sci. **305**, 408 (1994)

6.59    K. Ensslin, P.M. Petroff: Phys. Rev. B **41**, 12307 (1990)

6.60    K. Kern, D. Heitmann, P. Grambow, Y.H. Zhang, K. Ploog: Phys. Rev. Lett. **66**, 1618 (1991)

6.61    E. Zaremba: Phys. Rev. B **53**, 10512 (1996)

6.62    C. Dahl, F. Brinkop, A. Wixforth, J.P. Kotthaus, J.H. English, M. Sundaram: Solid State Commun. **80**, 673 (1991)

6.63    C. Dahl: Plasmaresonanzen in zweidimensionalen Elektronenscheiben auf GaAs. Fortschritt-Berichte (VDI, Düsseldorf 1993) Ser. 9, Vol. 165

6.64    C. Dahl, J.P. Kotthaus, H. Nickel, W. Schlapp: Phys. Rev. B **46**, 15590 (1992)

6.65    P. Bakshi, D.A. Broido, K. Kempa: J. Appl. Phys. **70**, 5150 (1991)

6.66    R. Landauer: In *Electrical Transport and Optical Properties of Inhomogeneous Media*, ed. by J.C. Garland, D.B. Tanner (AIP, New York 1978)

6.67    A.O. Govorov, A.V. Chaplik: J. Phys. Condens. Matter **6**, 6507 (1994)

6.68    E. Wigner: Phys. Rev. **46**, 1002 (1934)

6.69    C.C. Grimes, G. Adams: Phys. Rev. Lett. **42**, 795 (1979)

6.70    P.D. Tongaw, C.S. Lent: J. Appl. Phys. **75**, 1818 (1994)

6.71    R.A. Wyss, C.C. Eugster, J.A. del Alamo, Q. Hu, M.J. Rooks, M.R. Melloch: Appl. Phys. Lett. **66**, 1144 (1995)

6.72    L.P. Kouwenhoven, S. Jauhar, J. Orenstein, P.L. McEuen, Y. Nagamune, J. Motohisa, H. Sakaki: Phys. Rev. Lett. **73**, 3443 (1994)

6.73    R.C. Ashoori, H.L. Störmer, L.N. Pfeiffer, K.W. Baldwin, K. West: Phys. Rev. B **45**, 3894 (1992)

6.74    R.E. Collin: *Grundlagen der Mikrowellentechnik* (VEB Verlag Technik, Berlin 1973)

6.75    M. Wagner, A.V. Chaplik, U. Merkt: Phys. Rev. B **51**, 13817 (1995)

# Subject Index

# Topics in Applied Physics  Founded by Helmut K. V. Lotsch

Printing: Mercedesdruck, Berlin
Binding: Buchbinderei Lüderitz & Bauer, Berlin